Work Out

Physics

GCSE

The titles
in this
series

MACMILLAN
WORK OUT
SERIES

Work Out

Physics

GCSE

H. J. P. Keighley

MACMILLAN

First published 1986
Reprinted with corrections 1986
This edition 1987

Published by
MACMILLAN EDUCATION LTD
Houndmills, Basingstoke, Hampshire RG21 2XS
and London
Companies and representatives
throughout the world

Typeset by TecSet Ltd,
Sutton, Surrey
Printed and bound in Great Britain by Scotprint Ltd., Musselburgh, Scotland

British Library Cataloguing in Publication Data
Keighley, H. J. P.
Work out physics GCSE.—2nd ed.—
(Macmillan work out series)
1. Physics—Examinations, questions, etc.
I. Title   II. Keighley, H. J. P. Work out
physics
530'.076   QC32
ISBN 0-333-44004-8 (paper cover)
ISBN 0-333-36789-8 (export)

To Biz, for her patience, kindness and help

Contents

Preface

The book is designed to help students revise for the GCSE examination in physics. Each chapter contains complete answers to examination questions. These answers, together with the summaries at the beginning of each chapter, cover the work needed for the examinations of the different boards. Most students find it helpful to see and study worked examples, but it is, of course, essential that they understand them! To aid understanding, additional explanations have sometimes been added. Square brackets have been used for these explanations so that they are easily distinguished from the answer to the question. Cross-references are also given where it is felt that this will lead to a better understanding of the work.

In each chapter there is a section entitled 'Have you mastered the basics?' This is designed to enable students to ensure that they have mastered the essentials. The answers to these questions are given, in most cases, by reference to the appropriate section in the chapter.

Questions from past O-level papers which were in the previous edition have only been retained if they are typical of GCSE-type questions. Since many of the questions are actual examination questions, they sometimes include more than one topic. No attempt has been made to limit questions to ones on a single theme, as this would be unrealistic, but cross-references are given when it is felt that this would be helpful.

At the end of each chapter there is a selection of examination questions for the student to answer. Answers and considerable help with the solutions are provided. It is hoped that these will be an aid to students who do not have access to a tutor.

There is an index at the end of the book which may be used to revise a particular topic or a part of a topic.

Finally I must express my thanks to friends who have kindly read parts of the manuscript and given valuable advice.

Marlborough College, 1987   H. J. P. K.

Acknowledgements

The author and publishers wish to thank the Associated Examining Board, the University of London School Examinations Department, the University of Oxford Delegacy of Local Examinations, the Scottish Examination Board, the Southern Universities' Joint Board and the Welsh Joint Education Committee for permission to use questions from past examination papers.

Every effort has been made to trace all the copyright holders but if any have been inadvertently overlooked the publishers will be pleased to make the necessary arrangements at the first opportunity.

The University of London Entrance and School Examinations Council accepts no responsibility whatsoever for the accuracy or method in the answers given in this book to actual questions set by the London Board.

Acknowledgement is made to the Southern Universities' Joint Board for School Examinations for permission to use questions taken from their past papers but the Board is in no way responsible for answers that may be provided and they are solely the responsibility of the author.

The Associated Examining Board, the University of Oxford Delegacy of Local Examinations, the Northern Ireland Schools Examination Council and the Scottish Examination Board wish to point out that worked examples included in the text are entirely the responsibility of the author and have neither been provided nor approved by the Board.

Organisations Responsible for GCSE Examinations

In the United Kingdom, examinations are administered by the following organisations. Syllabuses and examination papers can be ordered from the addresses given here:

Northern Examining Association (NEA)

Joint Matriculation Board (JMB)
Publications available from:
John Sherratt & Son Ltd
78 Park Road, Altrincham
Cheshire WA14 5QQ

Yorkshire and Humberside Regional Examinations Board (YREB)
Scarsdale House
136 Derbyside Lane
Sheffield S8 8SE

North West Regional Examinations Board (NWREB)
Orbit House, Albert Street
Eccles, Manchester M30 0WL

North Regional Examinations Board
Wheatfield Road, Westerhope
Newcastle upon Tyne NE5 5JZ

Associated Lancashire Schools Examining Board
12 Harter Street
Manchester M1 6HL

Midland Examining Group (MEG)

**University of Cambridge Local
   Examinations Syndicate (UCLES)**
Syndicate Buildings, Hills Road
Cambridge CB1 2EU

**Oxford and Cambridge Schools
   Examination Board (O & C)**
10 Trumpington Street
Cambridge CB2 1QB

Southern Universities' Joint Board (SUJB)
Cotham Road
Bristol BS6 6DD

**East Midland Regional Examinations
   Board (EMREB)**
Robins Wood House, Robins Wood Road
Aspley, Nottingham NG8 3NR

**West Midlands Examinations Board
   (WMEB)**
Norfolk House, Smallbrook
Queensway, Birmingham B5 4NJ

London and East Anglian Group (LEAG)

**University of London School
   Examinations Board (L)**
University of London Publications Office
52 Gordon Square
London WC1E 6EE

**London Regional Examining Board
   (LREB)**
Lyon House
104 Wandsworth High Street
London SW18 4LF

East Anglian Examinations Board (EAEB)
The Lindens, Lexden Road
Colchester, Essex CO3 3RL

Southern Examining Group (SEG)

The Associated Examining Board (AEB)
Stag Hill House
Guildford, Surrey GU2 5XJ

**University of Oxford Delegacy of
   Local Examinations (OLE)**
Ewert Place, Banbury Road
Summertown, Oxford OX2 7BZ

**Southern Regional Examinations
   Board (SREB)**
Avondale House, 33 Carlton Crescent
Southampton, Hants SO9 4YL

**South-East Regional Examinations
   Board (SEREB)**
Beloe House, 2–10 Mount Ephraim Road
Royal Tunbridge Wells, Kent TN1 1EU

**South-Western Examinations Board
   (SWEB)**
23–29 Marsh Street
Bristol BS1 4BP

Scottish Examination Board (SEB)
Publications available from:
Robert Gibson and Sons (Glasgow) Ltd
17 Fitzroy Place, Glasgow G3 7SF

Welsh Joint Education Committee (WJEC)

245 Western Avenue
Cardiff CF5 2YX

**Northern Ireland Schools Examinations
  Council** (NISEC)

Examinations Office
Beechill House, Beechill Road
Belfast BT8 4RS

Introduction

How to Use this Book

This book is intended to help you prepare for your GCSE examination in Physics. It is not intended to replace a textbook but to supplement it. It is written to help you revise and prepare for an examination and assumes that you have completed a GCSE course. If there are topics that you have never studied, refer to a good textbook and carefully read and make notes on the relevant chapter. One good, concise and easy-to-read text for this purpose is *Mastering Physics*, published by Macmillan Education.

At the beginning of each chapter there is a pithy summary of the essential physics that you need to know. Spend time in becoming thoroughly familiar with the material in these summaries.

Each chapter contains GCSE-type questions with worked examples. Carefully studying these will help you to revise as well as showing you how to set out your answers in an examination. At the end of the worked examples there is a section entitled 'Have you mastered the basics?' Do these questions without looking at the solutions provided. Make sure you really grasp the principles and the basic physics contained in these questions. Finally there are questions for you to answer. Answers and fairly extended hints on solutions are provided for you if you get stuck.

Revision

When revising, many people find it helpful to write as they work. Writing as you revise will help you to concentrate and also help you to remember what you have been revising. You are much more likely to remember something you have written than something that you just looked at in a book. In particular, make sure that when you have revised a section of the work you have understood the basic laws, definitions and equations, and can use them.

You will probably find you can concentrate better and learn faster if you revise hard for a number of fairly short sessions rather than for one long one. Most people find that they get more revision done in four half-hour sessions, with breaks between them, rather than in one two-hour session.

Your memory recall of the work you have revised will be improved immensely if you look through it at regular intervals. Suppose you do an hour's revision. If you spend just five minutes revising the same work again the next day and another five minutes a few days later, your memory recall at some future date will be very greatly improved. The important point to remember about revision is that it should occur regularly for short periods rather than for one long period; also that work should be reviewed briefly shortly after you have done it and again a few days later.

A quick revision may be carried out by:

(i) learning thoroughly the summaries at the beginning of each chapter, and

(ii) answering the questions in 'Have you mastered the basics?', which are in each chapter at the end of the Examples. *Do them without looking at the solutions*!

A quick way of revising many of the basic principles is to go through the book answering the multiple choice questions.

To enable you to revise a particular topic there is an index at the end of the book.

The questions at the end of each chapter are to give you practice at answering examination questions.

Check your own syllabus carefully. Topics which only occur in some syllabuses have, wherever possible, been placed towards the end of the worked examples, so that they may be omitted easily. Syllabuses and past papers may be obtained from the addresses shown in the Acknowledgements.

How to Tackle Different Types of Examination Questions

(a) Calculations

You will not obtain full marks for a calculation unless you start by stating the physical principles involved and show the steps by which you arrive at the answer. You will lose marks in an examination if your answers to calculations consist of numbers without any indication of the reasoning by which you arrive at them. Remember it is a physics examination and not an arithmetic examination! The physics of the question must be clearly stated. Unless the answer is a ratio it will have a unit, so you *must always remember to state the unit*.

You may find it helpful before starting a calculation to ask yourself:

1. What am I being asked to calculate?
2. Which principle, law or formula needs to be used?
3. What units am I going to use?

A further hint worth remembering is that an apparently complex problem involving many items of data may often be clarified using a simple diagram.

A numerical answer must not have more significant figures than any number used in the calculation (see section 1.3).

Example of a Calculation

A *uniform plank of wood 3 m long is pivoted at its mid-point and used as a seesaw. Jean, who weighs 400 N, sits on one end. Where must John, who weighs 600 N, sit if the seesaw is to balance?*

To answer this question first draw a diagram showing the plank and the forces acting on it.

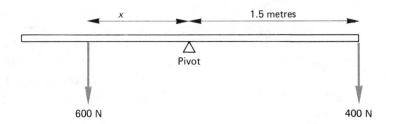

All the relevant information is shown on the diagram. We need to calculate the distance x metre so that the seesaw is balanced. The physical principle must first be stated, i.e. for a body in equilibrium:

Anticlockwise moment about a point = Clockwise moment about the same point

We now substitute in this equation, taking moments about the pivot

$$(600 \text{ N} \times x) = (400 \text{ N} \times 1.5 \text{ m})$$

$$\therefore x = \frac{400 \text{ N} \times 1.5 \text{ m}}{600 \text{ N}}$$

$$\therefore x = 1 \text{ m}$$

John must sit 1 m from the mid-point on the opposite side to Jean.

Always check your answer at the end to ensure that it is physically reasonable. Had you got an answer to the above question of 3 m, then you should at once realise that this is unreasonable (John would be off the end of the seesaw!), and look back to discover where you have made a mistake. *And don't forget to put a unit after the answer.*

(b) Multiple Choice Questions

If all the questions have to be attempted, do not waste valuable time reading through the paper before you start. Start at Question 1 and work steadily through the paper. If you come to a difficult question which you can't answer, miss it out and come back to it at the end. If you spend a lot of time thinking about questions which you don't find particularly easy, you may find yourself short of time to do some easier questions which are at the end of the paper.

When possible it is helpful to try and answer the question without first looking at the responses. This can reduce the chance of 'jumping to the wrong conclusions'. When you have found the response you think is correct, if time allows, check the other responses to see why they are incorrect.

When you have worked through the paper in this way, return to the questions you found difficult the first time through, but leave until the very end any questions about which you have little or no idea.

As examination boards do not deduct marks for wrong answers it is essential that you answer every question. If you are not sure which of the choices is the correct answer try and eliminate one or two of the alternative answers, and, if necessary, guess which of the remaining alternatives is correct.

If you have to alter an answer, ensure that the previous one has been completely rubbed out. Only one answer for each question must appear on the answer sheet.

Example of a Multiple Choice Question

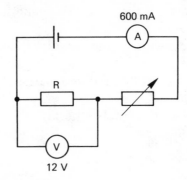

In the circuit shown in the diagram the high-resistance voltmeter reads 12.0 V, and the ammeter reads 600 mA. The value of the resistance marked R is

A 7200 Ω   **B** 72 Ω   **C** 20 Ω   **D** 7.2 Ω   **E** 0.02 Ω

Answer

We use the equation in Section 12.1 which defines resistance, namely

$$R = \frac{\text{potential difference across object (volts)}}{\text{current flowing in object (amperes)}}$$

The ammeter reading is 600 mA = 0.6 A

$$R = \frac{12 \text{ V}}{0.6 \text{ A}} = 20 \text{ Ω}$$

Answer **C**

(c) **Short Answer or Structured Questions**

In this type of question most examining boards leave space for the answer to be written on the question paper. The amount of space left will be a guide to the length of answer that is required, as will the number of marks indicated by the question. It doesn't follow that if you don't fill all the space up you haven't answered the question correctly. But if five lines are left for the answer, and you have only written on one line, or if you can't possibly get your answer on five lines, then you certainly ought to have another think about the answer.

Example of a Short Answer or Structured Question

A system of pulleys is used to raise a load of 9 N through 2 m. The effort of 2 N needed to do this moves through 12 m. What is
  (i) the potential energy gained by the load,
 (ii) the work done by the effort,
(iii) the efficiency of the system?

Answer

  (i) Work done on load = force x distance = 9 N x 2 m = 18 J
      Potential energy gained by load = 18 J

4

(ii) Work done by effort = force × distance = 2 N × 12 m = <u>24 J</u>

(iii) Efficiency = $\dfrac{\text{work out}}{\text{work in}}$ = $\dfrac{18\text{ J}}{24\text{ J}}$ = $\dfrac{3}{4}$ = <u>0.75 or 75%</u>

(If you have problems understanding the answer, refer to Chapter 4.)

(d) Free Response Type Questions

Make sure you attempt the full number of questions you have to answer.

If you are asked to describe an experiment you must normally draw a diagram of the apparatus or the relevant circuit diagram. A labelled diagram saves time, as information shown on the diagram need not be repeated in words. You should state clearly exactly what readings are taken, remembering that readings should be repeated as a check whenever possible. Finally mention any precautions necessary to obtain an accurate result.

Avoid saying vaguely at the end 'the result is calculated from the readings'. You must state exactly how it is calculated.

If the answer involves drawing a graph, be sure to label the axes and choose a scale so that the graph covers most of the paper.

Example of a Free Response Type Question

(a) Describe an experiment you would perform in order to measure the average power a girl can develop over a period of a few seconds. **(8 marks)**

(b) A car of weight 6000 N climbs a hill 1 km long which raises the car a vertical distance of 50 m. The driver maintains a constant speed of 25 m/s while he travels up the hill.

    (i) How long does it take him to reach the top of the hill? **(2 marks)**

    (ii) What is the gain in gravitational potential energy of the car when it reaches the top of the hill? **(4 marks)**

    (iii) Neglecting frictional forces, what is the power developed by the car as it climbs the hill? **(2 marks)**

(c) Outline the *main* energy changes which take place as the car climbs the hill. **(4 marks)**

Answer

(a) Leg muscle power may be measured by running up a flight of steps and measuring the time it takes with a stopwatch. Start the watch as the girl starts to climb the stairs and stop it when she reaches the top. Suppose that the average of three runs is 6 s. Measure the total vertical height of the stairs by using a ruler to measure the average height of each step and multiply by the number of steps to get the total vertical height. If this height is 5 m and the weight of the person, found by using bathroom scales, is 720 N, then

work done in 6 s = 720 N × 5 m = 3600 J

power = $\dfrac{\text{work done}}{\text{time taken}}$ [see Section 4.2] = $\dfrac{3600\text{ J}}{6\text{ s}}$ = 600 W

[The important thing in answering this question is to state clearly what must be measured: (1) the weight of the person, using bathroom scales; (2) the vertical height of the staircase, using a metre rule; (3) the time to climb the stairs, using a clock, taking the average number of runs. You must also show how to calculate her power.]

(b) (i) The car travels 1 km (1000 m) at 25 m/s; hence,

$$\text{time taken} = \frac{1000 \text{ m}}{25 \text{ m/s}} = 40 \text{ s}$$

(ii) Gravitational potential energy gained = weight × vertical height raised
[see Sections 4.2 and 4.4]   = 6000 N × 50 m = 300 000 J

(iii) power = $\dfrac{\text{work done}}{\text{time taken}}$ [see Section 4.2] = $\dfrac{300\,000 \text{ J}}{40 \text{ s}}$ = 7500 W

(c) The chemical energy of the fuel becomes heat energy in the engine cylinders, which eventually becomes gravitational potential energy of the car together with heat energy in the atmosphere as a result of frictional forces.

(e) Practical Course Work

Some points to bear in mind are:

(i) Follow any instructions carefully step by step. Read each instruction twice to ensure that you have understood it.

(ii) Take great care in reading an instrument. Record each reading accurately, carefully, systematically and clearly. When reading a horizontal scale with a pointer above it, make sure your eye is vertically above the pointer as you take the reading. Don't forget to put the UNIT. Whenever you read a scale, make sure you have carefully thought through the value of each of the small divisions on the scale. Remember to correct for zero errors (for example, make sure a spring balance reads zero before you hang a weight from it). When reading multiscale instruments, be careful to read the correct scale.

(iii) Present your readings clearly. Tabulate them neatly (don't forget the UNIT) and present them graphically whenever possible. Choose a suitable scale for the graph (i.e. 1 large square = 1 unit) and don't forget to label the axes. Take great care over plotting the points. Don't join them with short straight lines but draw the best smooth curve or straight line through them.

(iv) Repeat all the readings and take the average. Be particularly careful to check any unexpected reading.

(v) If you have to take the temperature of a liquid which is being heated, you must stir the liquid well before taking the temperature.

(vi) Be aware of the accuracy of any readings. For example, when reading a ruler marked in centimetres, you can only read it to the nearest 0.1 cm.

(vii) It is important to select the most appropriate apparatus. For example, if you need to measure a current of about 0.009 A (9 mA) and you have the choice of two ammeters, one reading 0–10 mA and the other 0–1 A, you should choose the 0–10 mA one.

(viii) You will need to know how to use, among other things, a measuring cylinder, a balance for measuring mass, a spring balance, a stopwatch, a thermometer, a voltmeter, an ammeter and a CRO.

(ix) Be prepared to criticise your work. State your conclusion clearly, and whenever possible state a concise relationship between variables.

A Few Important Points

(i) Read the question carefully before you write anything. Make sure you know exactly what the question is asking.

(ii) Answer the question precisely as asked. Note carefully the phrase used in the question. For example, 'explain', 'define', 'state', 'derive' and 'describe' all mean different things and are meant to be taken literally.

(iii) Always attempt the full number of questions specified even if you cannot answer them fully. It is much easier to obtain the first few marks of a new question than the last few marks of a previous question. The candidate who only answers four questions when he should answer five reduces his maximum mark to 80%. So keep an eye on the clock and make sure you do not spend too much time on any one question.

(iv) Set your work out neatly and clearly. An examiner is human and if he has done many hours of marking, he is likely to be far more sympathetic if your work is well set out and easy to follow.

(v) Make sure you are familiar with the style of question set by your particular examining board and the length of time allowed for each question (syllabuses and past papers may be obtained from the addresses given in the Acknowledgements).

(vi) Don't spend a lot of time on a Multiple Choice Question or a Short Answer Question with which you are having difficulty. Leave that question and come back to it later if you have time.

1 Some Help with Mathematics

1.1 Using Mathematical Equations

Equations help us to relate certain quantities. The usefulness of the equation is extended if it can be rearranged. There are three helpful rules for rearranging an equation.

(a) The Plus/Minus Rule

A symbol or number may be moved from one side of an equation to the other provided the sign in front of the symbol or number is changed. That is, a 'plus' item on one side becomes a 'minus' item on the other side and vice versa. For example:

$X = Y + 30$

$X - Y = 30$ or $X - 30 = Y$

(b) The Diagonal Rule

An item may be moved diagonally across an equals sign. For example, if

$$\frac{A}{B} = \frac{C}{D}$$

then the arrows show possible moves, so that

$$\frac{A}{B \times C} = \frac{1}{D} \quad \text{or} \quad \frac{A}{C} = \frac{B}{D} \quad \text{or} \quad \frac{D}{B} = \frac{C}{A}$$

(c) The 'Do unto Others' Rule

Whatever is done on one side of the equation must be done to the other side. For example:

if $A = B$ then $A^2 = B^2$ (both sides have been squared)

if $C = \dfrac{1}{D}$ then $\dfrac{1}{C} = D$ (both sides have been inverted)

BUT if $\dfrac{1}{R} = \dfrac{1}{R_1} + \dfrac{1}{R_2}$ then $R \neq R_1 + R_2$ ($\neq$ means 'does not equal')

The WHOLE of each side must be inverted, i.e.

$$R = \frac{1}{\dfrac{1}{R_1} + \dfrac{1}{R_2}} = \frac{1}{\dfrac{R_2 + R_1}{R_1 R_2}} = \frac{R_1 R_2}{R_1 + R_2}$$

1.2 Graphs

Choose easy scales, such as one large square to represent 1, 2 or 5 units (or multiples of 10 of these numbers). Avoid scales where one large square is 3, 4, 6, 7, 8 or 9 units. If each large square is, say, 3 units, then a small square is 0.3 units and this makes the plotting of the graph much more difficult. Usually the spread of readings along the two axes should be about the same and they should cover most of the page.

It is important to remember to label the axes and to put a title at the top of the page. In Physics most relationships are either straight lines or smooth curves, so it is not correct to join adjacent points together by short straight lines. You must decide whether the relationship is a straight line or a curve, and then draw either the best straight line through the points or a smooth curve.

1.3 Significant Figures

A useful rule to remember is that you must never give more significant figures in the answer than the number of significant figures given in the least precise piece of data. For example, if a cube of side 2.0 cm has a mass of 71.213 g, then we may calculate the density by using the equation

$$\text{density} = \frac{\text{mass}}{\text{volume}} \quad \text{[see Section 2.3]}$$

$$= \frac{71.213 \text{ g}}{8.0 \text{ cm}^3} = 8.901\ 625 \text{ g/cm}^3$$

according to my calculator! But the side was only given to two significant figures and we may not give the answer to more than two significant figures, i.e. 8.9 g/cm³ (we certainly do not know the density to the nearest millionth of a g/cm³!). The answer should be written 8.9 g/cm³.

A problem arises when the examiners use whole numbers without a decimal point and expect you to take them as exact. In the above calculation, if the side were given as 2 cm we may only strictly give the answer to one significant figure, i.e. 9 g/cm³. If you are in doubt in such a calculation, work the calculation to, say, 3 sig. figs. and then give the answer to 1 sig. fig.

1.4 Powers of Ten

100 may be written as 10^2. 1000 = 10^3

$\dfrac{1}{10}$ may be written as 10^{-1}.

$$\frac{1}{100} = 10^{-2} \qquad \frac{1}{1000} = 10^{-3} \qquad \frac{1}{100\,000} = 10^{-5}$$

1.5   Conversion of Units

$1 \text{ m}^2 = 100 \text{ cm} \times 100 \text{ cm} = 10^4 \text{ cm}^2$

so $5 \text{ N/cm}^2 = 5 \times 10^4 \text{ N/m}^2 = 50 \text{ kN/m}^2$

$1 \text{ m}^3 = 100 \text{ cm} \times 100 \text{ cm} \times 100 \text{ cm} = 10^6 \text{ cm}^3$

so $1 \text{ g/cm}^3 = 10^6 \text{ g/m}^3 = 1000 \text{ kg/m}^3$

Therefore to convert g/cm^3 to kg/m^3 we multiply by 1000.

1.6   Solidus and Negative Index Notation

Make sure you are familiar with the notation used by your examination board. Most boards use the solidus for GCSE examinations, i.e. m/s and kg/m^3. But m/s may be written m s^{-1} and kg/m^3 may be written kg m^{-3}.

2 SI Units, Density, Pressure and Hooke's Law

2.1 SI Units

(a) Some SI Base Units

Some base units of the SI system (Système International) are shown in Table 2.1.

Table 2.1

Physical quantity	Name of unit	Symbol for unit
Length	metre	m
Mass	kilogram	kg
Time	second	s
Current	ampere	A
Temperature	kelvin	K

(b) Prefixes

Some of the more commonly used prefixes are given in Table 2.2.

Table 2.2

Prefix	Sub-multiple	Symbol	Prefix	Multiple	Symbol
centi-	10^{-2}	c	kilo-	10^{3}	k
milli-	10^{-3}	m	mega-	10^{6}	M
micro-	10^{-6}	μ	giga-	10^{9}	G
nano-	10^{-9}	n			
pico-	10^{-12}	p			

2.2 Weight and Mass

The mass of a body (measured in kg) is constant wherever the body is situated in the Universe. The weight of a body (measured in N) is the pull or force of gravity on the body and this does depend on where the body is situated in the Universe.

The Earth's gravitational field is 10 N/kg , so the weight of a mass of 1 kg is 10 N. The pull of the Earth (weight) on an apple of average size is about 1 N.

2.3 Density

$$\text{Density} = \frac{\text{mass}}{\text{volume}}$$

If the mass is in kg and the volume in m^3, then the density is in kg/m^3.

Density may be determined by measuring the mass of a measured volume.

2.4 Pressure

$$\text{Pressure} = \frac{\text{normal force}}{\text{area}}$$

If the force is measured in newtons and the area in $(\text{metre})^2$, then the pressure is in pascals (Pa). 1 Pa = 1 N/m^2.

The pressure due to a column of liquid (i) acts equally in all directions, (ii) depends on the depth and the density of the liquid. It may be calculated using the equation

pressure (Pa) = 10 (N/kg) × depth (m) × density (kg/m^3)

which may also be written as

pressure (Pa) = $\rho g h$

where ρ is the density in kg/m^3, g the acceleration due to gravity (10 m/s^2) and h the depth in metres.

Pressure may be measured using a U-tube manometer or a Bourdon gauge. When any part of a confined liquid is subject to a pressure, the pressure is transmitted equally to all parts of the vessel containing the liquid. This principle, and the fact that liquids are virtually incompressible, are made use of in hydraulic machines. Such machines are useful force multipliers. Referring to Fig. 2.1 the pressure on the small piston is 20 N/10 cm^2 or 2 N/cm^2. This pressure is transmitted to the large piston and the force on it is 2 $\dfrac{N}{cm^2}$ × 100 cm^2 = 200 N.

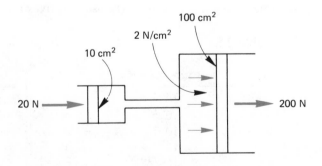

Figure 2.1 The principle of hydraulic machines.

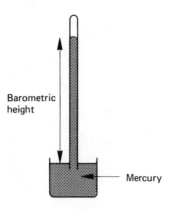

Figure 2.2 A mercury barometer. The atmosphere exerts a force on the surface of the mercury in the trough and this pushes the mercury up the tube.

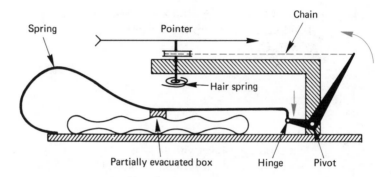

Figure 2.3 An aneroid barometer. When the atmospheric pressure increases, the centre of the partially evacuated box moves inwards and this small movement is magnified by a system of levers. The chain attached to the end lever moves the pointer. The large spring prevents the box from collapsing.

The atmosphere above us exerts a pressure known as the atmospheric pressure. If the air is withdrawn from a metal can, the force due to atmospheric pressure acting on the outside of the can will collapse the can. It is because of the decrease in atmospheric pressure with height that aircraft cabins have to be pressurised. The atmospheric pressure may be measured using a mercury barometer (Fig. 2.2) or an aneroid barometer (Fig. 2.3).

2.5 Hooke's Law

Hooke's law states that provided loads are not used which would cause a spring or wire to approach its *elastic limit*, the extension is proportional to the applied load.

Until the elastic limit is reached a spring (or wire) returns to its original length if the load is removed.

2.6 Worked Examples

Example 2.1

A mass of 1 kg is secured to the hook of a spring balance calibrated on the Earth. The spring balance reading is observed when it is freely suspended at rest just above the Earth's surface, secondly inside a spaceship orbiting round the Earth, and finally at rest on the Moon's surface.

If the acceleration due to free fall on the Earth is $10 \, \text{m/s}^2$ and acceleration due to free fall on the Moon is $1.6 \, \text{m/s}^2$, the spring balance readings, in N, would be (Table 2.3):

Table 2.3

	Point above Earth's surface	Inside a spaceship	On the Moon
A	1.0	0	0.16
B	1.0	0.84	0.16
C	10.0	0	1.6
D	10.0	0.84	0.16
E	10.0	11.6	1.6

(AEB)

Solution 2.1

[Weight on Earth = 1 kg × 10 N/kg = 10 N.
Weight on Moon = 1 kg × 1.6 N/kg = 1.6 N.
Inside the spaceship the weight is zero.]

Answer **C**

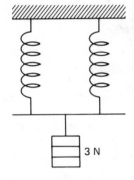

3 N

Example 2.2

A student applies a force of 6 N to a helical spring and it extends by 12 cm. He then hangs the spring in parallel with an identical spring and attaches a load of 3 N as shown. The resulting extension of the system, in cm, will be

A 3   **B** 4   **C** 6   **D** 12   **E** 24

(AEB)

Solution 2.2

[Force acting on end of each spring = 1.5 N (the total upward force must equal the
total downward force)

6 N extends the spring by 12 cm. Assuming Hooke's law applies (Section 2.5),

1.5 N extends the spring by $\left(\dfrac{12}{6} \times 1.5\right)$cm   = 3 cm.]

Answer **A**

Example 2.3

A U-tube containing mercury is used as a manometer to measure the pressure of gas in a container. When the manometer has been connected, and the tap opened, the mercury in the U-tube settles as shown in the diagram.

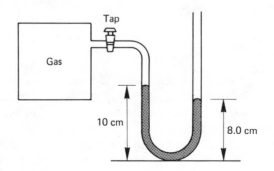

The pressure of the atmosphere is equal to that exerted by a column of mercury of length 76 cm. The pressure of the gas in the container is equal to that exerted by a column of mercury of length

      A 2.0 cm   **B** 58 cm   **C** 74 cm   **D** 78 cm   **E** 94 cm

Solution 2.3

[The atmospheric pressure exerted on the open limb of the U-tube is greater than the gas pressure by 2 cm of mercury. The gas pressure is therefore $76 - 2 = 74$ cm.]

Answer C

Example 2.4

    (a) A block of stone measures 2 m × 2 m × 1 m. It has a mass of 8000 kg.
      (i) What is its density?
      (ii) When it is standing on a bench what is the maximum pressure it can exert on the bench?
    (b) A bath has some water in it and the depth of the water at the shallow end is 0.2 m. At the plug hole end it is 0.3 m. What pressure does the water exert on the plug?
    (The Earth's gravitational field is 10 N/kg and the density of water is 1000 kg/m^3.)

                                                          (10 marks)

Solution 2.4

(a) (i) Density = $\dfrac{\text{mass}}{\text{volume}}$ [Section 2.3] $= \dfrac{8000 \text{ kg}}{4 \text{ m}^3} = \underline{2000 \text{ kg/m}^3}$

    (ii) Maximum pressure occurs when the area of contact is a minimum. Minimum area in contact with bench = 2 m^2.

        Pressure = $\dfrac{\text{force}}{\text{area}}$ [Section 2.4] $= \dfrac{80\,000 \text{ N}}{2 \text{ m}^2} = 40\,000 \text{ Pa} = \underline{40 \text{ kPa}}$

(b) Pressure = $10\ \dfrac{\text{N}}{\text{kg}}$ × depth × density [Section 2.4]

     = $10\ \dfrac{\text{N}}{\text{kg}}$ × 0.3 m × $1000\ \dfrac{\text{kg}}{\text{m}^3} = \underline{3000 \text{ Pa}}$

[The pressure on the plug depends on the vertical height of the water above the plug.]

Example 2.5

A vehicle designed for carrying heavy loads across mud has four wide low-pressure tyres, each of which is 120 cm wide. When the vehicle and its load have a combined mass of 12 000 kg each tyre flattens so that 50 cm of tyre is in contact with the mud as shown in the diagram.

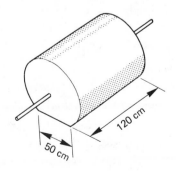

(a) Calculate
    (i) the total area of contact of the vehicle tyres with the mud,
    (ii) the pressure exerted on the mud.
(b) A car of mass 1000 kg is unable to travel across the mud although it is much lighter than the load-carrying vehicle. Why is this?

(6 marks)

(L)

Solution 2.5

(a) Area = 4 × (120 × 50) cm² = 24 000 cm²

(b) Pressure = $\dfrac{force}{area}$ = $\dfrac{(12\,000 \times 10)\,N}{24\,000\ cm^2}$ = 5 N/cm² or 50 kN/m²

(c) The area of car tyre in contact with the road is much less than that of the vehicle. The pressure exerted by the car tyre on the mud is therefore greater than that exerted by the vehicle, and the tyre sinks into the mud.

Example 2.6

> Describe how you would obtain, as accurately as possible, a series of readings for the load and corresponding extension of a spiral spring.   **(6 marks)**
> A student obtained the following readings:

Load/N	0	1	2	3	4	5	6
Length of spring/cm	10.0	11.5	13.0	14.5	16.0	18.5	24.0

> Using these results, plot a graph of load against extension and estimate the load beyond which Hooke's law is no longer obeyed.   **(7 marks)**
> The spring is at rest with a mass of 0.2 kg on its lower end. It is then further extended by a finger exerting a vertical force of 0.5 N. Draw a diagram showing the forces acting on the mass in this position, giving the values of the forces.   **(3 marks)**
> Describe the motion of the mass when the finger is removed. Make your description as precise as possible, by giving distances. State the position where the kinetic energy of the mass will be greatest.   **(4 marks)**

(L)

Solution 2.6

Clamp the top of the spring firmly to a support, making sure that the support is also firm and cannot move. Clamp a ruler alongside the spring and attach a horizontal pointer to the bottom of the spring in such a way that the pointer is close to the surface of the ruler (this will help to avoid a parallax error when taking the readings). Record the pointer reading. Hang a known load on the end of the spring and again record the pointer reading. Increase the load and record the new pointer reading. Continue in this way, thus obtaining a series of readings. The extension is calculated for each load by subtracting the unloaded pointer reading from the loaded pointer reading. A check may be made by again recording the pointer readings as the loads are removed one at a time (this is also a means of checking that the elastic limit has not been reached).

[Graph (Fig. 2.4). Remember to label axes, choose suitable scales, and when the line is no longer straight, draw a smooth curve. You are asked to plot extension against load, so the original length of the spring, 10 cm, must be subtracted from each reading.]

Hooke's law is obeyed for loads up to 4 N but very soon after this the graph begins to curve and Hooke's law is no longer obeyed.

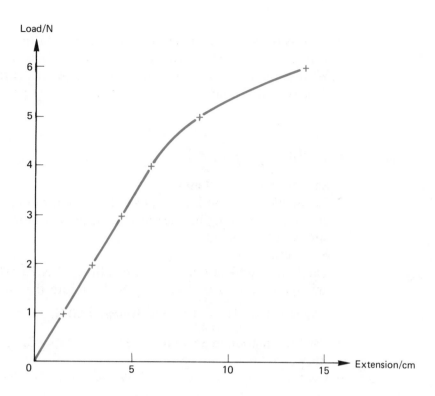

Figure 2.4 Load against extension.

The 0.2 kg mass (force 2 N) extends the spring 3 cm to a position of equilibrium. A further force of 0.5 N extends the spring 0.75 cm beyond this position. When the 0.5 N force is removed the spring oscillates about the original position

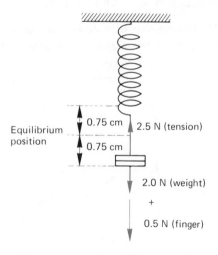

of equilibrium. The oscillations will gradually decrease from an amplitude of 0.75 cm to zero. The spring will then be at rest with an extension of 3 cm. The maximum kinetic energy is when the spring passes through the equilibrium position (extension 3 cm).

2.7 Have You Mastered the Basics?

1. Can you define density, pressure and elastic limit?
2. Can you state Hooke's law and describe how to obtain force–extension graphs for rubber, nylon, copper and springs?
3. Can you describe a mercury barometer, the principle physics in the working of an aneroid barometer, and a hydraulic machine?

4. What is the weight of a mass of 5 kg which is on the Earth?
5. A force of 12 N acts on an area of 3 m x 2 m. What is the pressure?
6. What is the hydrostatic pressure 20 m below the surface of a lake of water of density 1000 kg/m³?

2.8 Answers and Hints on Solutions to 'Have You Mastered the Basics?'

1. See Sections 2.3, 2.4 and 2.5.
2. See Section 2.5 and Example 2.6. For nylon or copper it is best to clamp one end, attach a spring balance or weights hung over pulleys to the other end, and stretch it horizontally.
3. See Section 2.4.
4. The Earth's gravitational field strength is 10 N/kg. This means that the pull of gravity on a mass of 1 kg is 10 N. Therefore the pull of gravity on a mass of 5 kg is $(5 \text{ kg}) \left(10 \dfrac{\text{N}}{\text{kg}}\right) = 50$ N. Weight = 50 N.
5. Use the equation in Section 2.4. The area is (3×2) m². Answer = 2 N/m² = 2 Pa.
6. Use the equation in Section 2.4. Pressure = 2×10^5 Pa.

2.9 Questions

Question 2.1

(Answers and hints on solutions will be found in Section 2.10.)

The gravitational field strength on the surface of the Moon is 1.6 N/kg. Which of the pairs of values shown below applies to a mass taken to the Moon?

	Mass (in kg)	Weight (in N)
A	10	0.16
B	10	1.6
C	10	16
D	1.6	1.0
E	1.6	16

Question 2.2

(a) Define density. State a consistent set of units in which the quantities could be measured.
(b) An empty box has a mass of 2 kg and is made from material which has a density of 8000 kg/m³. What is the volume of material which is needed to make the box?
(c) The density of a liquid is 1200 kg/m³. What is the liquid pressure 8 m below the surface of the liquid. Take the acceleration of free fall as 10 m/s². **(10 marks)**

Question 2.3

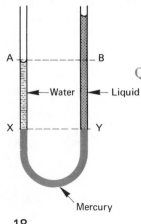

The U-tube shown in the diagram contains a liquid and water separated by mercury. Which of the following statements concerning the arrangement is correct?
A The pressure at Y is greater than the pressure at X
B The pressure at A is the same as the pressure at B
C The density of water is greater than the density of the liquid
D The pressure of the water column is greater than the pressure of the liquid column

18

Question 2.4

A ballroom floor can withstand a maximum pressure of 4000 kN/m² without damage. Which one of the following would damage it?

A A woman weighing 0.6 kN standing on the heel of one shoe of area 10^{-4} m²

B An elephant weighing 200 kN standing on one foot of area 0.1 m²

C A 1000 kN load standing on an area of 100 m²

D A man weighing 0.8 kN standing on both feet if the area of his feet in contact with the ground is 0.04 m²

Question 2.5

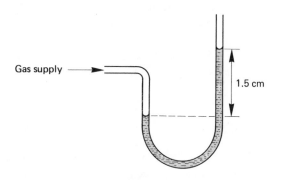

Gas supply

1.5 cm

An manometer containing mercury is connected to a gas supply as shown. If atmospheric pressure is equivalent to 750 mm mercury, the pressure of the gas in mm of mercury is

A 600   **B** 700   **C** 735   **D** 765   **E** 900   **(AEB)**

Question 2.6

(a) Explain what is meant by the phrase 'pressure at a point in a liquid'.   **(3 marks)**

(b) The diagram illustrates part of the braking system of a motor car. The master cylinder and its piston are connected hydraulically to a slave cylinder and its piston.

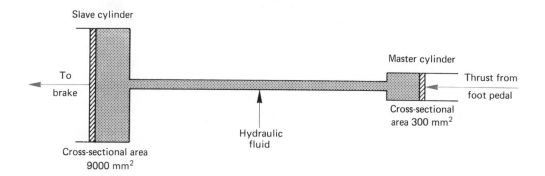

Slave cylinder

Master cylinder

To brake

Thrust from foot pedal

Cross-sectional area 300 mm²

Hydraulic fluid

Cross-sectional area 9000 mm²

If the thrust on the piston in the master cylinder is 600 N, use the information in the diagram to calculate

(i) the pressure in the hydraulic fluid,

(ii) the thrust on the piston in the slave cylinder.   **(6 marks)**

(c) State one property of a liquid that makes it suitable as a hydraulic fluid.   **(1 mark)**

Question 2.7

(a) What is meant by 'pressure'? **(2 marks)**

(b) State two factors on which the pressure exerted by a fluid depends. **(2 marks)**

(c) What is the pressure of the atmosphere in pascals on a day when a mercury barometer reads 760 mmHg? (The density of mercury is 13 600 kg/m^3 and the Earth's gravitational field is 10 N/kg.) **(4 marks)**

Question 2.8

(a) Describe an experiment you would perform to investigate the relationship between the extension of a spring and the load hung on it. Your answer should include a sketch of the apparatus, a description of all the observations you would make and how you would deduce the relationship from your readings. **(12 marks)**

(b) Distinguish between mass and weight. Which does a spring balance read? Explain why a sensitive spring balance can detect the variation in the Earth's gravitational field as it is taken to different parts of the Earth's surface, but a sensitive beam balance is unable to detect any variation. **(8 marks)**

Question 2.9

The diagram shows three identical spring balances X, Y and Z, supporting a light rod to which is attached a weight of 60 N. X and Z are equidistant from Y.

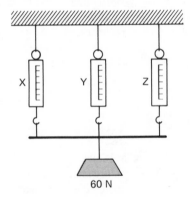

60 N

The table shows possible readings of the spring balances in newtons. Which set of readings is correct?

	X	Y	Z
A	60	60	60
B	20	60	20
C	20	20	20
D	10	40	10
E	0	60	0

(L)

20

2.10 Answers and Hints on Solutions to Questions

1. The force on 1 kg is 1.6 N; hence, the force on 10 kg is 16 N.
   Answer **C**
2. (a) See Section 2.3. (b) Use equation in Section 2.3.
   Answer 2.5×10^{-4} m^3
   (c) See Section 2.4.
   Answer 9.6×10^4 Pa
3. Answer **C**
4. Use the definition of pressure in Section 2.4. For woman, pressure = 6000 kN/m^2. The elephant, the load and the man exert pressures of less than 4000 kN/m^2.
   Answer **A**
5. Gas pressure = atmospheric pressure + pressure due to 15 mm of mercury.
   Answer **D**
6. (a) Take a small area round the point and work out the pressure from pressure = force/area.
   (b) (i) 2 N/mm^2 or 2×10^6 Pa. (ii) Remember that the pressure is the same throughout the fluid.
   Answer 18 000 N
   (c) See Section 2.4.
7. (a) and (b) see Section 2.4. (c) Use equation in Section 2.4. What units must the height of the mercury column be in?
   Answer 1.03×10^5 Pa
8. (a) There are 12 marks for this part, so full experimental details are needed. Remember to clamp the spring in a support which will not move when the weights are hung from the spring. Check the elastic limit is not exceeded by checking the readings when the weights are removed. Plot load against extension and a straight line through the origin shows that the extension is proportional to the load.
   (b) See Section 2.2. A beam balance compares masses and the mass does not change with position on the Earth's surface. The extension of a spring balance depends on the pull of gravity on the mass hung on its end.
9. Answer **C**

3 Motion, Scalars and Vectors

3.1 Velocity and Acceleration

$$\text{Velocity} = \frac{\text{distance travelled in a specified direction (i.e. displacement)}}{\text{time taken}}$$

(Unit m/s)

$$\text{Acceleration} = \frac{\text{change in velocity}}{\text{time taken for change}} \qquad \text{(Unit m/s}^2\text{)}$$

The gradient of a displacement–time graph is the velocity. The gradient of a velocity–time graph is the acceleration. The area under a velocity–time graph is the distance travelled (N.B. the velocity axis must start from zero).

3.2 Newton's Laws of Motion

1. If a body is at rest, it will remain at rest; and if it is in motion, it will continue to move in a straight line with a constant velocity unless it is acted on by a resultant external force.
2. The acceleration of a body is directly proportional to the resultant force acting on it and inversely proportional to the mass of the body.
3. If a body A exerts a force on a body B, then B exerts an equal and opposite force on body A.

Newton's second law may be verified using a trolley and ticker timer as described in Example 3.8. The second law may be summarised by the equation

$F = ma$

where F is the force in newtons, m the mass in kilograms and a the acceleration in metres per second per second.

3.3 Scalars and Vectors

A *scalar* quantity has magnitude only. A *vector* quantity has magnitude and direction. Examples of scalar quantities are mass, temperature, speed and energy. Force, weight and velocity are examples of vector quantities. Notice that velocity is a vector quantity, whereas speed is a scalar. A body travelling at 2 m/s on a circular path has a constant speed but its velocity is changing.

Vector quantities must be added by the rule for vector addition. Forces of 3 N and 4 N acting at right angles have a resultant of 5 N. This may be verified by drawing a diagram (Fig. 3.1) to scale and measuring the length of the resultant R. Alternatively R may be calculated using the measurements shown in the diagram.

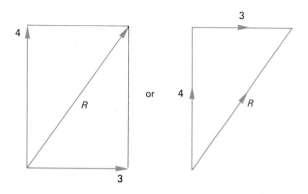

Figure 3.1 Addition of vectors.

3.4 Uniformly Accelerated Motion

If a body has a velocity u and accelerates with an acceleration a, travels for a time t, reaching a velocity v after travelling a distance s, then

$$\text{average velocity} = \frac{u + v}{2} = \frac{s}{t}$$

$$\text{distance travelled } (s) = (\text{average velocity}) \times t$$

$$\text{acceleration } (a) = \frac{v - u}{t}$$

$$\text{velocity } (v) = at$$

A body falling freely in a vacuum has a uniform acceleration known as the acceleration due to gravity or the acceleration of free fall. If the body is falling in air, eventually the forces due to air resistance are equal and opposite to the force of gravity on the body. When this state is reached, the body travels with a constant velocity known as the *terminal velocity*. (Two examples are parachutes and raindrops.)

The acceleration due to gravity (the acceleration in free fall) may be measured using a ticker-timer (see Example 3.8) or by using photodiodes and a centisecond clock (see Question 3.7).

3.5 Worked Examples

Example 3.1

The acceleration, in m/s^2, of the body whose motion is represented by the graph shown is
A 1.5   B 2.0   C 3.5   D 6.0   E 14.0   (AEB)

23

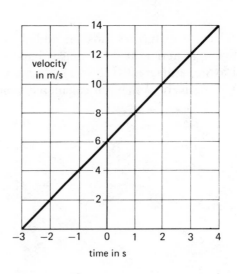

Solution 3.1

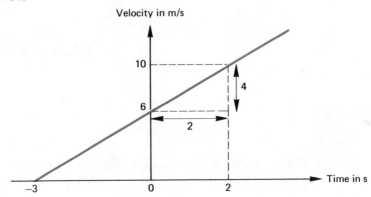

$$\left[\text{Acceleration} = \frac{\text{change in velocity}}{\text{time taken for change}} = \text{gradient of graph (see Section 3.1)} \right.$$

$$\left. = \frac{4\text{ m/s}}{2\text{ s}} = 2\text{ m/s}^2 \right]$$

Answer **B**

Example 3.2

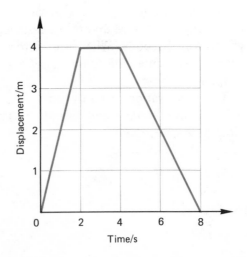

The graph shows how the displacement of a body varies with the time.
(a) Describe the motion of the body.   (4 marks)
(b) What is the velocity of the body during the first 2 s?   (3 marks)
(c) How far does the body travel in the first 4 s?   (2 marks)

24

Solution 3.2

(a) The displacement at time $t = 0$ s is zero. The body then moves with a constant velocity until it has moved 4 m, and this takes 2 s. From 2 s to 4 s the body is stationary. It then returns to its original position at a constant velocity, and this velocity is half the original velocity.

(b) Velocity = $\dfrac{\text{distance travelled}}{\text{time taken}} = \dfrac{4 \text{ m}}{2 \text{ s}} = 2$ m/s

(c) The body travels 4 m in the first 2 s and is then stationary for 2 s. Distance travelled is 4 m.

Example 3.3

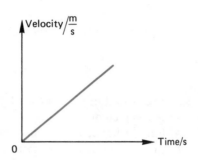

The graph shows how the velocity of an object varies with the time. The object is
A moving with a decreasing acceleration
B moving with a constant acceleration
C moving with an increasing acceleration
D moving with a constant velocity

Solution 3.3

[The increase in the velocity in a given time is always the same, i.e. the gradient of the graph is constant, so the acceleration is constant.]
<u>Answer **B**</u>

Example 3.4

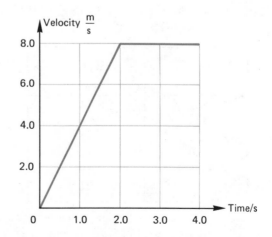

The graph is a velocity–time graph for a trolley.
(a) What is the acceleration of the trolley during the first 2 s?   **(3 marks)**
(b) What is the acceleration of the trolley between 2 s and 4 s?   **(1 mark)**
(c) How far does the trolley travel in 4 s?   **(4 marks)**

25

Solution 3.4

(a) Acceleration = $\dfrac{\text{change in velocity}}{\text{time taken for change}}$ [see Section 3.1]

$$= \frac{8.0 \text{ m/s}}{2.0 \text{ s}} = 4.0 \text{ m/s}^2$$

(b) Zero. [The velocity is constant.]

(c) Distance travelled = area under graph [see Section 3.1]

$$= (\text{area of triangle}) + (\text{area of rectangle})$$

$$= \left(\tfrac{1}{2} \times 2.0 \text{ s} \times 8.0 \ \frac{\text{m}}{\text{s}}\right) + \left(8.0 \ \frac{\text{m}}{\text{s}} \times 2.0 \text{ s}\right)$$

$$= 8.0 \text{ m} + 16.0 \text{ m} = 24.0 \text{ m}$$

Example 3.5

A force of 10 N gives a mass of 5 kg an acceleration of 2 m/s^2. The same force would produce an acceleration of

A 1 m/s^2 when acting on a mass of 15 kg
B 2 m/s^2 when acting on a mass of 10 kg
C 5 m/s^2 when acting on a mass of 2 kg
D 10 m/s^2 when acting on a mass of 2 kg

Solution 3.5

[The three quantities are related by the equation $F = ma$ (see Section 3.2). If $F = 10$ N then the product ma must be 10 N. This is true for response **C**.]

Answer **C**

Example 3.6

The table below shows the distances travelled by the average family car when it is driving at certain speeds, from the moment a hazard is spotted, until the brake is applied (thinking distance) and the distance to stop after the brake is applied (braking distance). The last column shows the total distance travelled from the moment the hazard is spotted until the car comes to rest.

Speed / $\dfrac{\text{km}}{\text{h}}$	Thinking distance/m	Braking distance/m	Overall distance/m
32	6	6	12
48	9	14	23
64	12	24	36
80	15	38	53
96	18	55	73
110	21	75	96

1. (a) What is the thinking distance for a speed of 32 km/h? **(1 mark)**
   (b) At a speed of 9 m/s (about 32 km/h), how long does it take the car to travel 18 m?

   (2 marks)

   (c) What is the 'thinking' *time* at a speed of 18 m/s (64 km/h)? **(3 marks)**
   (d) What is the relationship between speed and thinking distance? **(2 marks)**
   (e) What would you expect to be the thinking distance at 160 km/h? **(2 marks)**

2. (a) If the car decelerates uniformly during the braking, what is the *average* speed during the braking if the car starts at 18 m/s (64 km/h) and braking continues until the car comes to rest? **(2 marks)**

   (b) At this average speed, how long would it take to go 9 m? **(1 mark)**

   (c) At this average speed, (i) what is the braking distance? (ii) how long after the brakes were applied would the car take to stop? **(4 marks)**

3. Two similar cars are driving along the same road at 110 km/h. They are 90 m apart and travelling in the same direction. Use the distances shown in the table to decide whether this means that they will inevitably collide if the front car brakes suddenly and the rear car brakes as soon afterwards as he can? Explain your answer. **(3 marks)**

Solution 3.6

1. (a) 6 m.

   (b) 9 m in 1 s.
   ∴ 18 m in 2 s.

   (c) Thinking time = time to travel 12 m = $\dfrac{12 \text{ m}}{18 \text{ m/s}}$ = $\dfrac{2}{3}$ s = 0.67 s

   (d) They are proportional. [Doubling the speed doubles the thinking distance.]

   (e) 30 m. [It is twice the thinking distance at 80 km/h.]

2. (a) Average speed = $\dfrac{18 \text{ m/s}}{2}$ = 9 m/s

   (b) 1 s. [It is travelling at 9 m/s, so it takes 1 s to go 9 m.]

   (c) Braking distance is 24 m.

   ∴ Time to stop = $\dfrac{24 \text{ m}}{9 \text{ m/s}}$ = 2.7 s

   [The time to stop is the time to go 24 m at a speed of 9 m/s.]

3. No. After the front car brakes, it travels 75 m. The overall stopping distance of the second car is 96 m, so they will not crash and will be about (90 m − 21 m) = 69 m apart when they have both come to rest.

   [The cars are 90 m apart when the first car applies its brakes. The rear car travels 21 m before applying its brakes. *Both* travel a further 75 m before coming to rest.]

Example 3.7

Oil is leaking from a car as it travels along the road. One drop falls to the ground every 2 s. The diagram (not drawn to scale) shows the pattern of the drops on the road.

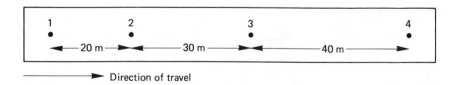

(a) Is the car accelerating, decelerating or travelling with a constant velocity? Give a reason for your answer. **(2 marks)**

(b) The distance on the road between the drops marked 1 and 2 is 20 m, the distance on the road between drops 2 and 3 is 30 m, and the distance on the road between drops 3 and 4 is 40 m. What is the acceleration of the car? **(6 marks)**

Solution 3.7

(a) The car is accelerating, because the distance between successive drops is increasing.

(b) Average velocity during first 2 s = $\dfrac{\text{distance}}{\text{time taken}}$ = $\dfrac{20\text{ m}}{2\text{ s}}$ = 10 m/s

The acceleration is constant and 10 m/s is the velocity at the midpoint − i.e. after 1 s.

Velocity after 1 s = 10 m/s

Velocity after 3 s = $\dfrac{30\text{ m}}{2\text{ s}}$ = 15 m/s

Acceleration = $\dfrac{\text{change in velocity}}{\text{time taken}}$ = $\dfrac{(15-10)\text{ m/s}}{2\text{ s}}$

[The velocity changes from 10 m/s to 15 m/s between 1 s and 3 s − i.e. in 2 s.]

$\Rightarrow$ Acceleration = 2.5 m/s²

Example 3.8

(a) In order to investigate the relationship between force and acceleration, an experiment was carried out using a ticker-tape attached to a trolley as shown in the diagram. The

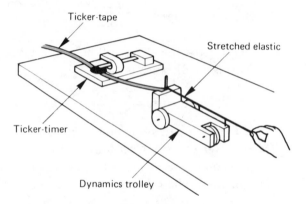

Ticker-tape

Stretched elastic

Ticker-timer

Dynamics trolley

tape was fed through a ticker-timer which made 50 dots on the tape every second. The trolley was placed on a gently sloping inclined plane. A force was applied by pulling on an elastic band attached to the trolley.

(i) Explain why an inclined plane was used and what experiment would have been conducted in order to get the correct inclination of the plane. **(3 marks)**

(ii) How would the force applied to the trolley have been kept constant? **(1 mark)**

(iii) How would a force which was twice the magnitude of the original force have been applied? **(1 mark)**

(b) The diagram below shows two sections of one tape obtained from such an experiment.

Dot 10   Dot 35

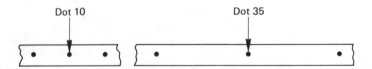

The first section shows dots 9, 10 and 11; the second section dots 34, 35 and 36.

(i) Measure the distance between dots 9 and 11 and calculate the average velocity of the trolley between dots 9 and 11. **(3 marks)**

(ii) Calculate the average velocity of the trolley between dots 34 and 36. **(2 marks)**

(iii) What time interval elapsed between dot 10 and dot 35? **(1 mark)**

(iv) Calculate the acceleration of the trolley. **(3 marks)**

(c) The acceleration of the trolley was calculated for five different forces.
　　(i) Sketch a graph which represents the results of the above experiment.　(**2 marks**)
　　(ii) State the relationship between force and acceleration.　(**2 marks**)

Solution 3.8

(a)　(i) The inclined plane is to compensate for friction. The plane is tilted until the component of the gravitational force accelerating the trolley is equal to the frictional force. This is done by tilting the plane until the trolley moves with a constant velocity when given a push. When the correct tilt is obtained, a ticker-tape attached to the trolley will have dots on it equally spaced.

　(ii) The rubber band must be kept stretched by the same amount throughout the run.

(iii) Two identical rubber bands each stretched the same amount as the original rubber band.

(b)　(i) Distance between dot 9 and dot 11 = 2 cm.
　　Time between dot 9 and dot 11 = $\frac{2}{50}$ s = 0.04 s.

$$\text{Velocity} = \frac{\text{distance gone}}{\text{time taken}} = \frac{2 \text{ cm}}{0.04 \text{ s}} = 50 \text{ cm/s}$$

　(ii) Distance between dot 34 and dot 36 = 5 cm.

$$\text{Velocity} = \frac{5 \text{ cm}}{0.04 \text{ s}} = 125 \text{ cm/s}$$

(iii) 0.5 s. [This is the time for 25 dots.]

(iv) Acceleration = $\dfrac{\text{change in velocity}}{\text{time taken for change}} = \dfrac{(125 - 50) \text{ cm/s}}{0.5 \text{ s}} = 150 \text{ cm/s}^2$

(c)　(i)

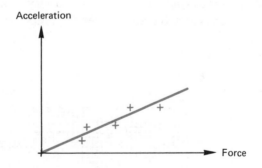

　(ii) The acceleration is proportional to the resultant force, provided the mass is kept constant.
　　[or Resultant force = mass × acceleration]

[The relationship between mass and acceleration may be investigated by keeping the force constant (band at constant stretch) and varying the mass (add masses to the trolley). The acceleration due to gravity, *g*, may be measured if the ticker-timer is arranged so that the tape (with a mass on its end) falls vertically.]

Example 3.9

　(a) What do Newton's laws tell us about the effect of a force on a body?　(**3 marks**)
　(b) Define a newton.　(**3 marks**)

(c) A body of mass 5 kg is at rest when a horizontal force is applied to it. The force varies with time as shown on the graph below. Use the figures on the graph to calculate how the velocity varies with time and plot a graph of velocity against time for the first 30 s of its motion. **(11 marks)**

(d) How far does it travel in the first 10 s? **(3 marks)**

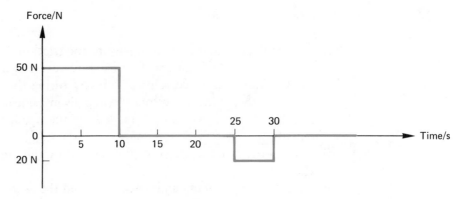

Solution 3.9

(a) When a resultant force acts on a body, the body accelerates – that is, changes its velocity. The acceleration is proportional to the force and inversely proportional to the mass.

(b) A newton is the force which gives a mass of 1 kg an acceleration of 1 m/s^2.

(c) $F = ma$. During the first 10 s, 50 N = 5 kg $\times a$; hence, a = 10 m/s^2. For the next 10 s no force acts and the body continues with constant velocity. From 25 s to 30 s the body decelerates, and $-$ 20 N = 5 kg $\times a$; hence, a = $-$ 4 m/s^2. Since $v = at$ [see Section 3.5], after 10 s velocity = 10 m/s^2 $\times$ 10 s = 100 m/s. From 10 s to 25 s the body continues at 100 m/s. Between 25 s and 30 s the change in velocity is 4 m/s^2 $\times$ 5 s = 20 m/s. So, after 30 s the velocity is (100 $-$ 20) m/s = 80 m/s.

(d) Distance travelled = average velocity $\times$ time = 50 $\times$ 10 = 500 m.
[This is also the area under the graph for the first 10 s.]

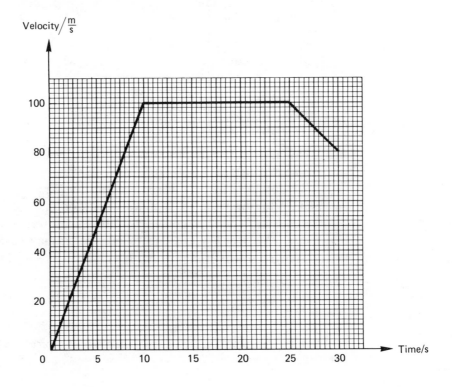

3.6 Have You Mastered the Basics?

1. Can you explain what is meant by velocity, acceleration, scalar and vector?
2. Do you know what the gradient of a displacement–time graph represents?
3. Do you know what the gradient of a velocity–time graph is and what the area under the graph represents?
4. Can you describe experiments to (i) show that the acceleration of a body is proportional to the force acting on it and inversely proportional to the mass of the body; (ii) measure the acceleration due to gravity?
5. Do you know the units of the quantities in the equation $F = ma$?
6. A force of 10 N acts on a mass of 2 kg. What is the acceleration?
7. Do you know how to add vectors?

3.7 Answers and Hints on Solutions to 'Have You Mastered the Basics'?

1. See Sections 3.1 and 3.3.
2. See Section 3.1.
3. See Section 3.1.
4. See Example 3.8 and Question 3.7.
5. See Section 3.2.
6. $F = ma \Rightarrow 10\ \text{N} = 2\ \text{kg} \times a \Rightarrow a = 5\ \text{m/s}^2$.
7. See Section 3.3.

3.8 Questions

(Answers and hints on solutions will be found in Section 3.9.)

Question 3.1

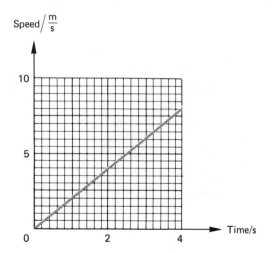

The graph shows the speed of a body of mass 5 kg moving in a straight line plotted against the time.

(a) Use the graph to calculate the acceleration of the mass.　　　　　　**(3 marks)**

(b) Use the graph to determine the distance travelled by the mass in the first 4.0 s.

　　　　　　(3 marks)

(c) What force acts on the body?　　　　　　**(3 marks)**

Question 3.2

(a) The graph shows the velocity of a lift plotted against time.
  (i) What is the acceleration of the lift during the first 2 s? **(2 marks)**
  (ii) Describe the motion of the lift between 2 s and 4 s. **(1 mark)**
  (iii) What is the total distance travelled by the lift during its journey? **(3 marks)**

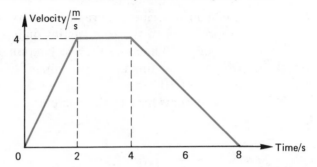

(b) In an experiment to measure g, the acceleration of free fall, a tape had a mass of 0.5 kg attached to its end. The mass was allowed to fall and, as it fell, it pulled the tape through a ticker-timer which made 50 dots on the tape every second. The diagram shows the tape (not drawn to scale).
  (i) Use the measurement shown on the diagram to calculate the velocity at X.
  (2 marks)
  (ii) The velocity at Y was calculated and found to be 215 cm/s. What is the change in velocity between X and Y? **(1 mark)**
  (iii) Calculate the acceleration of the 0.5 kg mass between X and Y. **(2 marks)**
  (iv) What result would you expect if the experiment were repeated with a 2 kg mass on the end of the tape? **(1 mark)**

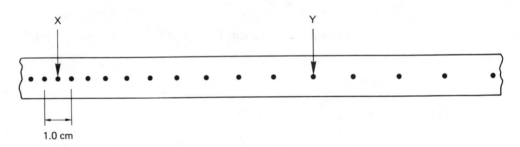

Question 3.3

A force of 10 N gives a mass of 5 kg an acceleration of 2 m/s^2. The same force would produce an acceleration of
A 4 m/s^2 when acting on a mass of 10 kg
B 4 m/s^2 when acting on a mass of 2.5 kg
C 10 m/s^2 when acting on a mass of 2 kg
D 5 m/s^2 when acting on a mass of 10 kg

Question 3.4

(a) What does Newton's first law of motion tell us about the effect of a force on a body?
  (3 marks)
(b) Sketch graphs to show the relationship between velocity and time for a trolley moving with (i) zero acceleration, (ii) an acceleration which increases with time. **(4 marks)**
(c) Describe, giving full details, how you would investigate the relationship between force, mass and acceleration. **(13 marks)**

Question 3.5

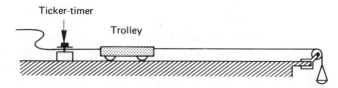

(a) The diagram shows a trolley on a horizontal bench. A scale pan is attached to one end of the trolley by means of a light string which runs over a pulley. A paper tape which runs through a ticker-timer is attached to the other end of the trolley.

(i) When a small mass is put into the pan and the trolley is given a gentle push, a series of dots is obtained on the tape as shown below. (Only every tenth dot is shown; the time interval between each dot shown in the diagram is $\frac{1}{5}$ s.)

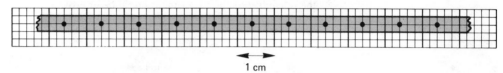

What conclusions can you draw from the dots? **(2 marks)**

(ii) An extra mass is put into the pan, the trolley is released and a series of dots is obtained as shown below. (Again only every tenth dot is shown.)

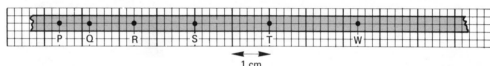

From this diagram
- (1) find the average speed, u, between P and Q, **(2 marks)**
- (2) find the average speed, v, between T and W, **(2 marks)**
- (3) show that the acceleration of the trolley is constant, **(2 marks)**
- (4) find the acceleration of the trolley, a. **(4 marks)**

(b)

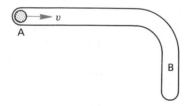

The diagram shows a horizontal table top, seen from above, with a smooth L-shaped groove in its surface.

A ball is projected from A and travels from A to B along the groove. What differences, if any, do you expect in
- (i) the speed of the ball at A and B, **(2 marks)**
- (ii) the velocity of the ball at A and B, **(2 marks)**
- (iii) the kinetic energy of the ball at A and B, **(2 marks)**

(Give a reason for each answer.)

(L, part question)

Question 3.6

(a) A ticker-timer makes 50 dots every second. A tape moving with constant velocity passes through the ticker-timer. The distance between dot 5 and dot 7 is 1 cm. What is the velocity of the tape?

(b) A body of mass 500 g has a resultant force of 2 N acting on it. What is its acceleration?

(c) A force acts for 3 s on a body and produces an acceleration of 2 m/s². Sketch the speed-time graph for the first 5 s of its motion. How far did it travel in the 5 s?

Question 3.7

(a) The diagram illustrates an experiment to measure g, the acceleration of free fall. A ball is dropped which breaks a light beam falling on a photodiode. The clock is started when the ball stops the light shining on the photodiode. Vertically below the first light beam a second light beam shines on another photodiode. When the ball continuing in its fall breaks this beam, the clock is stopped.

   (i) Mark on the diagram the distance you would measure. **(1 mark)**

  (ii) t (the time taken to travel a distance s) and s are related by the equation

$$s = \tfrac{1}{2}gt^2$$

     If s is 5 cm and t is 0.1 s, calculate the value of g. **(2 marks)**

(iii) Explain one way, other than taking the average of a number of readings, by which the reliability of the experiment could be improved. **(2 marks)**

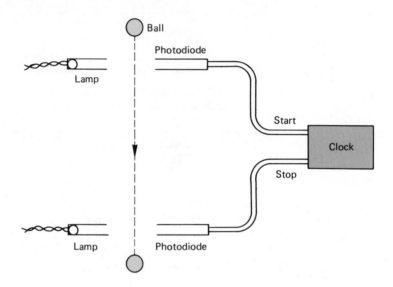

(b) In an experiment to measure the acceleration of free fall, a mass of 1 kg was attached to a piece of tape. The mass was allowed to fall freely and the tape passed through a ticker-timer which made 50 dots on the tape every second. The tape (not drawn to scale) is shown below. Measurements taken from the tape show that the velocity at X was 25 cm/s and the velocity at Y was 125 cm/s.

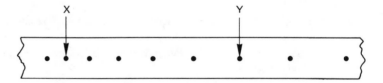

   (i) Calculate the acceleration of the 1 kg mass between X and Y. **(3 marks)**

  (ii) What would the result be if the experiment were repeated with a 2 kg mass on the end of the tape. **(1 mark)**

(iii) State one precaution you would take to ensure an accurate result. **(1 mark)**

Question 3.8

What is meant by (a) a scalar quantity, (b) a vector quantity? **(4 marks)**
What is the magnitude of the resultant vector when two forces each of magnitude 4 N act at an angle of 60° to each other? **(6 marks)**

3.9 Answers and Hints on Solutions to Questions

1. (a) Acceleration = $\dfrac{8 \text{ m/s}}{4 \text{ s}}$ = 2 m/s².

   (b) Distance travelled = area under graph = ($\frac{1}{2}$ x 8 x 4) m = 16 m.
   (c) $F = ma$ = 5 kg x 2 m/s² = 10 N.

2. (a) (i) Use acceleration = $\dfrac{\text{change in velocity}}{\text{time taken}}$ [see Section 3.1] = 2 m/s².

   (ii) The lift is moving at a constant velocity of 4 m/s. (iii) Distance travelled is the total area under the graph (see Section 3.1) = 4 m + 8 m + 8 m = 20 m.

   (b) (i) Velocity at X = 1.0 cm/0.04 s = 25 cm/s. (ii) 190 cm/s.
   (iii) $\dfrac{190 \text{ cm/s}}{0.2 \text{ s}}$ = 950 cm/s². (iv) The same.

3. Use $F = ma$. Product ma must equal 10 N.
   <u>Answer **B**</u>

4. (a) See Section 3.2.
   (b) See Fig. 3.2.

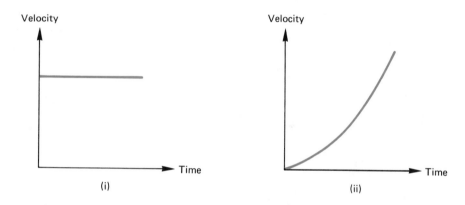

Figure 3.2

(c) The experiment is described in Example 3.8. The mass is kept constant to investigate the relationship between force and acceleration. The relationship between mass and acceleration is investigated by keeping the force constant (one rubber band at constant stretch) and varying the mass (this can be done by having two trolleys with one piled on the other, then three piled up, then four and five. A graph of acceleration against $1/m$ is a straight line through the origin.

5. (a) (i) The trolley is travelling with a constant velocity. (ii) (1) Average speed (u) = (distance travelled)/(time taken) = (0.8/0.2) cm/s = 4 cm/s. (2) (2.4/0.2) cm/s = 12 cm/s. (3) The acceleration is constant, because in equal time intervals the distance travelled always increases by the same amount: in this case 0.4 cm (2 squares). (4) Acceleration = $(v - u)$/time taken [see Section 3.4] = $(12 - 4)/(4 \times 0.2)$ cm/s² = 10 cm/s².

   (b) (i) No difference (because speed is a scalar quantity). (ii) The velocity has changed, because the direction of travel has changed and velocity is a vector quantity (see Section 3.3). (iii) Since the groove is smooth, we neglect friction, so energy is conserved and the kinetic energy at A is equal to the kinetic energy at B.

6. (a) Tape travels 1 cm in 1/25th of a second. Velocity = 25 cm/s.
   (b) Use $F = ma \Rightarrow a = 4$ m/s^2.
   (c) For the first 3 s the graph is a straight line, reaching 6 m/s after 3 s. It is then horizontal. Distance = area under graph (see Section 3.1) = 9 m + 12 m = 21 m.

7. (a) (i) Measure the distance between the two photodiodes. (ii) 5 cm = $\frac{1}{2} g$ (0.1 s)^2 $\Rightarrow g = 1000$ cm/s^2 or 10 m/s^2. (iii) Increase s. This will increase the accuracy of the measurement of s and the time interval t.
   (b) (i) The time interval between dots X and Y is 5 intervals of $\frac{1}{50}$ s = 0.1 s. Use acceleration = (change in velocity)/(time taken) = (125 cm/s − 25 cm/s)/(0.1 s) = 1000 cm/s^2. (ii) The same because all masses fall with the same acceleration. (iii) The tape should be arranged so as to run as freely as possible in order to reduce frictional forces.

8. See Section 3.3. Draw the vector diagram and measure the magnitude and direction of the resultant.
   Answer 6.9(3) N at 30° to each force

4 Moments, Equilibrium, Work, Energy, Power and Machines

4.1 Moments and Equilibrium

(a) Equilibrium and the Principle of Moments

When a body is in equilibrium, the resultant force on it is zero and the sum of the clockwise moments about a point is equal to the sum of the anticlockwise moments about the same point. In Fig. 4.1 a beam is balanced at its centre point. Using the measurements shown in the diagram,

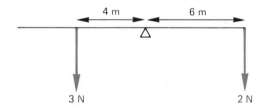

Figure 4.1 Clockwise moment = Anticlockwise moment = 12 N m.

clockwise moment = 2 N x 6 m = 12 N m
anticlockwise moment = 3 N x 4 m = 12 N m

The clockwise moment equals the anticlockwise moment and the beam is in equilibrium. In an experiment the forces are usually applied by hanging weights on the beam.

(b) Bending of Beams

Beams bend when placed under stress. In Fig. 4.2 the top of the beam is in compression and the underside is in tension. A force that causes stretching is called a tensile force. All parts of a stressed beam are in equilibrium under the action of the internal forces between neighbouring atoms. The greater the cross-sectional area of the beam, the greater the resistance to bending under stress.

Whenever a body changes shape by bending or stretching or compressing, it is said to be *elastic* if it returns to its original shape or length when the stress is removed. It is said to be *plastic* if it stays in its new shape or if it stays in its stretched position.

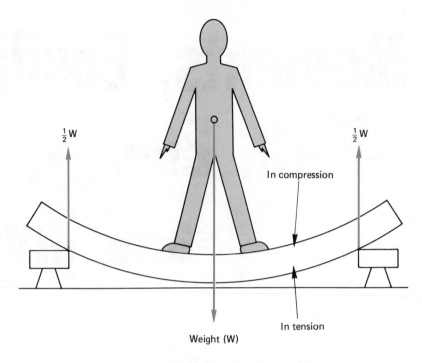

Figure 4.2 A man of weight **W** standing on the middle of a plank. The plank sags because the plank is not perfectly stiff. The top of it will shorten and is in compression. The underside will get longer and so is in tension.

(c) Centre of Gravity (Centre of Mass)

The centre of gravity of a body is the point through which its whole weight may be considered to act. (Determination of this is described in Example 4.12.)

(d) Types of Equilibrium

A body is said to be in stable equilibrium if when given a small displacement and then released it returns to its original position.

A body is said to be in unstable equilibrium if when given a small displacement and then released it moves further from its original position.

A body is said to be in neutral equilibrium if when given a small displacement and then released it stays in its new position.

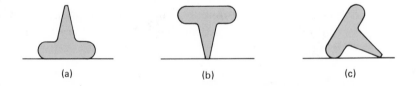

Figure 4.3 (a) Stable equilibrium, (b) unstable equilibrium, (c) neutral equilibrium.

Referring to Fig. 4.3, the equilibrium is stable in (a), unstable in (b) and neutral in (c). For a body in stable equilibrium, the stability is increased by having a large base area and/or a low centre of gravity. The object in Fig. 4.4(a) will return to its original position flat on the table when released, but in the situation shown in Fig 4.4(b) the object will fall when released. The stance adopted in karate ensures a strong position: the position of the feet gives in effect a large base area and the bending of the knees lowers the centre of gravity.

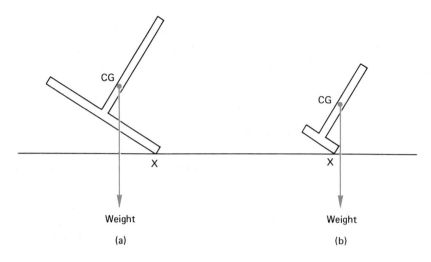

Figure 4.4 In (a) the moment about X of the weight tends to turn the body anticlockwise. In (b) the moment about X tends to turn the body clockwise and the body falls over.

4.2 Work, Energy and Power

(a) *Work* is energy transfer, and may be calculated from the equation work = force × distance moved in direction of force. Unit, joule (J). 1 *joule* of work is done when a force of 1 N moves its point of application through a distance of 1 m.

(b) *Energy* is the capacity to do work. It is measured in joules. The principle of conservation of energy states that energy cannot be created or destroyed, although it can be changed from one form to another.

  (i) *Potential energy* is energy possessed by a body by virtue of its position or the state the body is in. If a body is lifted vertically, then it gains gravitational potential energy. The gain in gravitational potential energy may be calculated from the equation

  gravitational potential energy gained = (weight) × (vertical height raised)

  Other examples of bodies possessing potential energy are the energy stored in a stretched rubber band, and the energy stored in electric and magnetic fields.

  (ii) *Kinetic energy* is energy possessed by a moving body: for a body of mass m moving with velocity v, the kinetic energy is $\frac{1}{2}mv^2$. If a body of mass m is lifted through a height h, then the work done is the force (mg) multiplied by the distance moved (h), which is mgh. The body has potential energy mgh (g is the Earth's gravitational field strength, which is 10 N/kg, or the acceleration of free fall, which is 10 m/s^2). If the mass is now released and it falls, then its potential energy changes into kinetic energy. When it has fallen the distance h and has a velocity v, then $mgh = \frac{1}{2}mv^2$.

  (iii) *Sources of energy.* Some of the available sources of energy are as follows. (a) *Fossil fuels* such as *coal, diesel fuel* and *petroleum* (the known sources of these portable fuels are being gradually exhausted). (b) *Wind, waves, hydroelectric, solar, biomass* (from plants – e.g. sugar is turned into ethanol; the main advantage is that it is a renewable source; the main problem is the amount of land needed to grow sufficient sugar cane). The origin of these sources of energy is the Sun. (c) *Tidal.* The origin is due mainly to the relative motion of the Moon and Earth. (d) *Geothermal.* The origin is heating due to radioactive decay in the interior of the Earth.

(iv) *Energy conversions in power stations.* Energy is obtained from (a) fossil fuels, (b) nuclear power, (c) hydroelectric power. In power stations the energy obtained from these sources is used to drive an electric generator. In (a) and (b) the energy is used to produce steam to drive the turbine which drives the generator. In (c) the gravitational potential energy of the water is turned into kinetic energy of the water and the moving water drives the turbine which drives the generator.

In pumped storage systems, during low periods of demand for electrical energy, water is pumped to the top of a hill, giving it gravitational potential energy. Later, when it flows down the hill, the gravitational potential energy becomes kinetic energy. The kinetic energy drives the turbine generators and electrical energy is produced. Heat energy results from most conversion processes, because friction always produces heat energy.

The block diagram below shows the principal energy changes which take place in a power station.

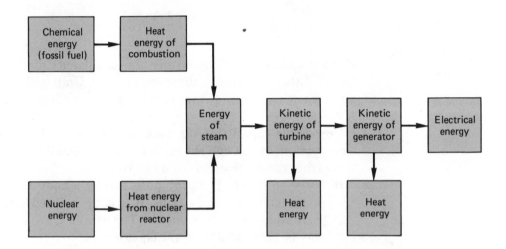

(v) *Batteries* provide transportable electrical energy. Rechargeable batteries are more expensive than non-rechargeable ones, but because they can be recharged they are cheaper in the long run. Their internal resistance (the electrical resistance of the chemicals inside them) is small and they are rapidly discharged and damaged if short-circuited. Their low internal resistance means that less energy is lost as heat in driving currents through the chemicals in the battery. They last less time before running down than ordinary batteries and they run down by themselves when not in use faster than non-rechargeable batteries.

(c) *Power* is the rate of doing work, or power $= \dfrac{\text{work done}}{\text{time taken}} =$ force × velocity.

Unit, watt (W).

1 *watt* (W) is a rate of working of 1 joule per second (J/s).

4.3 Friction

Friction is the force tending to prevent one body sliding over another. When two bodies are in contact and at rest, the maximum frictional force between them is known as the limiting static friction. When sliding occurs the force is known as the sliding friction. Friction may be reduced by the use of lubricants and ball bearings.

4.4 Machines

(a) Efficiency of a Machine

$$\text{Efficiency} = \frac{\text{work got out of a machine}}{\text{work put into a machine}} = \frac{\text{power output}}{\text{power input}}$$

Efficiency is a ratio and has no units.

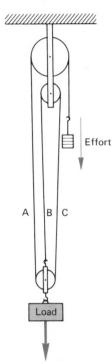

(b) Pulleys

For the machine shown in Fig. 4.5, the work got out of the machine is equal to the potential energy gained by the load and the work put into the machine is the work done by the effort. If the load rises 1 m then each of the strings A, B and C must shorten by 1 m and the effort moves 3 m.

(c) The Inclined Plane

When the load is lifted a vertical height h (Fig. 4.6) and the effort moves along the plane,

$$\text{efficiency} = \frac{\text{work got out}}{\text{work put in}} = \frac{\text{energy gained by load}}{\text{work done on load}}$$

$$= \frac{\text{weight} \times \text{vertical height raised}}{\text{effort} \times \text{distance effort moves}} = \frac{mgh}{\text{effort} \times d}$$

[Remember that the gravitational potential energy gained by the load is the weight multiplied by the vertical height raised.]

Figure 4.5 A simple pulley system.

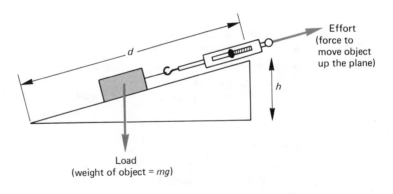

Figure 4.6 An inclined plane.

(d) Gears

The ratio between the number of rotations of each of the two intermeshed gears is given by

$$\frac{\text{number of rotations of driving gear}}{\text{number of rotations of driven gear}} = \frac{\text{number of teeth on driven gear}}{\text{number of teeth on driving gear}}$$

(e) Wheel and Axle

The steering wheel on a car provides a good example of the wheel and axle principle. In the absence of the steering wheel it would not be possible to steer the car. You could not provide a strong enough force. The steering wheel enables the force applied to it to be magnified. It increases the distance of the force from the centre of the axle and, hence, increases the turning moment.

(f) Heat Engines

In a petrol engine the mixture of petrol and air is ignited and heat energy is produced. The force of the resulting explosion gives the piston kinetic energy. The burnt gases are discharged at a lower temperature. Thus, chemical energy produces heat energy which is turned into kinetic energy of the piston.

4.5 Worked Examples

(When needed, take the Earth's gravitational field strength as 10 N/kg.)

Example 4.1

Name the SI unit of (i) energy; (ii) work; (iii) force; (iv) mass; (v) resistance.

Solution 4.1

(i) joule; (ii) joule; (iii) newton; (iv) kilogram; (v) ohm.

Example 4.2

Energy may be defined as
A producing power
B causing motion
C ability to do work
D ability to exert a force

Solution 4.2

Answer **C** [see Section 4.2]

Example 4.3

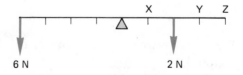

The diagram illustrates a uniform beam pivoted at its centre. The marks on the beam show equal distances each of 1 m in length. Loads of 6 N and 2 N are hung from the positions shown. Which of the following additional loads would keep the beam in equilibrium?

A 20 N at X

B 6 N at Y

C 7 N at Y

D 4 N at Z

Solution 4.3

Total anticlockwise moment about pivot = 6 N x 4 m = 24 N m. Thus, total clockwise moment must be 24 N m.

The 2 N force provides a clockwise moment of 2 N x 2 m = 4 N m, so the additional load must provide a moment of (24 − 4) N m = 20 N m.

The additional moments are:

A 20 N x 1 m = 20 N m; **B** 6 N x 3 m = 18 N m; **C** 7 N x 3 m = 21 N m; **D** 4 N x 4 m = 16 N m

Answer **A**

Example 4.4

(a) What power is produced by a machine which lifts a mass of 2 kg through a vertical height of 10 m in 2 s?

(b) A mass of 3 kg is thrown vertically upwards with a kinetic energy of 600 J. To what height will it rise?

(6 marks)

Solution 4.4

(a) Force needed to lift 2 kg = 20 N.

$$\text{Power} = \frac{\text{work done}}{\text{time taken}} = \frac{20 \times 10}{2} \text{ J/s} = 100 \text{ W}$$

(b) 600 J = *mgh*. [See Section 4.2.]

$$\therefore 600 \text{ J} = (3 \text{ kg}) \left(10 \, \frac{\text{N}}{\text{kg}}\right) h$$

$\therefore h = 20$ m

Height = 20 m

Example 4.5

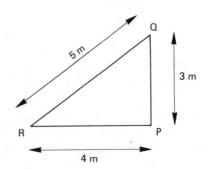

A weight of 5 N is moved up a frictionless inclined plane from R to Q as shown. What is the work done in joules?

A 15   **B** 20   **C** 25   **D** 35   **E** 60

(AEB)

Solution 4.5

Work done = weight × vertical height raised [see Section 4.2]
    = 5 N × 3 m
    = 15 J

<u>Answer A</u>

Example 4.6

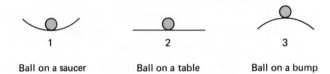

	1	2	3
A	Stable	Unstable	Neutral
B	Unstable	Neutral	Stable
C	Stable	Neutral	Unstable
D	Neutral	Stable	Unstable
E	Unstable	Stable	Neutral

The diagrams show a ball resting on different-shaped surfaces. They represent three different types of equilibrium, which are

Solution 4.6

[Use the summary in Section 4.1(d).]

<u>Answer C</u>

Example 4.7

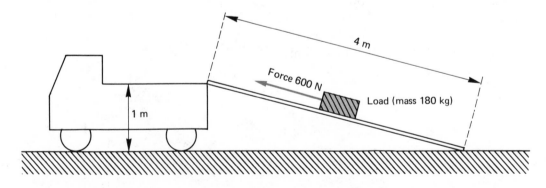

The diagram shows a ramp being used to get a load, which has a mass of 180 kg, onto a lorry. The ramp is 4 m long and the end of the lorry is 1 m above the ground. A force of 600 N is needed to pull the load up the ramp.
(a) Calculate
    (i) the gravitational potential energy gained by the load as it goes from the bottom to the top of the ramp; **(3 marks)**
    (ii) the work done by the 600 N force in pulling the load up the ramp; **(2 marks)**
    (iii) the efficiency of the system. **(2 marks)**

(b) As the lorry starts, the load topples off and falls to the ground. What is the kinetic energy of the load just before it hits the ground? **(2 marks)**

(Take the Earth's gravitational field strength as 10 N/kg.)

Solution 4.7

(a) (i) Gravitational potential energy gained = *mgh* [see Section 4.2]

$$= (180 \text{ kg})\left(10 \ \frac{\text{N}}{\text{kg}}\right)(1 \text{ m}) = 1800 \text{ J}$$

(ii) Work done = force × distance [see Section 4.2] = 600 N × 4 m = 2400 J

(iii) Efficiency = $\dfrac{\text{work out}}{\text{work in}}$ [see Section 4.4] = $\dfrac{1800 \text{ J}}{2400 \text{ J}} = \dfrac{3}{4} = 75\%$

[The gravitational potential energy gained is equal to the work done against gravity, and a force of 1800 N is lifted vertically through 1 m, so 1800 J of work are done, and this is the work got out of the machine.]

(b) Just before the load hits the ground all the gravitational potential energy will have turned into kinetic energy. Kinetic energy = 1800 J.

Example 4.8

A person who has a mass of 50 kg runs up some stairs in 9 s. The stairs are 8 m high. His power output is

A $\dfrac{50 \times 8}{9}$ W   B $\dfrac{50 \times 9}{8}$ W   C $\dfrac{50 \times 10 \times 8}{9}$ W   D $\dfrac{50 \times 10 \times 9}{8}$ W

Solution 4.8

[The Earth's gravitational field strength is 10 N/kg and the weight of the man is therefore $(50 \text{ kg})\left(10 \ \dfrac{\text{N}}{\text{kg}}\right) = (50 \times 10) \text{ N}$

Work done = force × distance [see Section 4.2]

$= (50 \times 10 \text{ N})(8 \text{ m}) = (50 \times 10 \times 8) \text{ J}$

Power = $\dfrac{\text{work done}}{\text{time taken}} = \dfrac{(50 \times 10 \times 8) \text{ J}}{9 \text{ s}} = \dfrac{(50 \times 10 \times 8)}{9}$ W

Hence, **C** is the correct answer.

<u>Answer **C**</u>

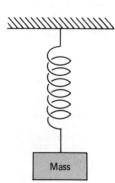

Example 4.9

A spring is fixed to a support at one end and a mass is hung on the other end. The mass is raised until the spring just goes slack. It is then released. From the moment the mass is released until it reaches its lowest point describe the changes that take place in

(i) the kinetic energy of the mass   **(3 marks)**

(ii) the gravitational potential energy of the mass   **(2 marks)**

(iii) the energy stored in the spring.   **(2 marks)**

Solution 4.9

(i) The kinetic energy is initially zero. (Remember that kinetic energy is energy possessed by a *moving* body.) The kinetic energy increases until the spring passes through its equilibrium position, when the kinetic energy is a maximum, and then the kinetic energy decreases until it reaches zero when the spring is at its lowest position.

(ii) The gravitational potential energy is a maximum and it gradually decreases until the spring reaches its lowest position.

(iii) This is initially zero and increases until it is a maximum when the spring is at its greatest extension.

Example 4.10

(a) Explain what is meant by (i) work; (ii) power.   **(4 marks)**

(b) The table below shows the power needed to bicycle up a hill (a) on a bicycle with upright handlebars and (b) with a lightweight bicycle with dropped handlebars. Study the table carefully and then answer the following questions:

Power needed to bicycle up hill

Type of bicycle	Speed/$\frac{m}{s}$	Power to overcome resistance/W	Extra power needed to climb hill/W	Total power/W
Upright handlebars	2.2	15	200	215
	4.4	52	400	452
	8.8	300	800	1100
Lightweight dropped handlebars	2.2	8	185	193
	4.4	30	370	400
	8.8	165	740	905

(i) What is the relationship between the extra power needed to climb up the hill and the speed? Explain your answer.   **(3 marks)**

(ii) Does the same relationship exist between the power to overcome resistance and speed? Explain your answer.   **(2 marks)**

(c) During a period of about half an hour the average cyclist can produce energy at about 220 W. Would you be able to climb the hill at (i) 2 m/s on both kinds of bicycle; (ii) 5 m/s on both kinds of bicycle? Explain your answers.   **(4 marks)**

(d) Draw a block diagram showing the energy changes which take place as the cyclist cycles up the hill.   **(4 marks)**

(e) Using a given amount of power, a bicycle with dropped handlebars enables you to bicycle up the hill at a greater speed than using a bicycle with upright handlebars. State one reason, other than the fact that it is lighter, why this is so.   **(1 mark)**

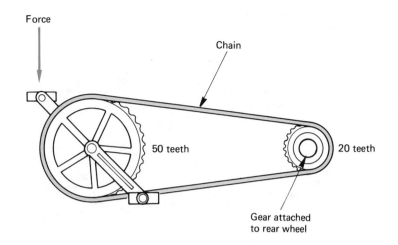

Force

Chain

50 teeth

20 teeth

Gear attached
to rear wheel

(f) The chain of a bicycle runs round two gears as shown in the diagram. The gear attached to the rear wheel has 20 teeth, and the gear attached to the pedals has 50 teeth. If the pedals go through two complete revolutions, how many times does the rear wheel rotate? **(2 marks)**

Solution 4.10

(a) (i) Work is done whenever a force moves through a distance. It is calculated from the equation

work done (J) = force (N) × distance moved in the direction of the force (m)

   (ii) Power is rate of doing work and may be calculated from the equation

$$\text{power} = \frac{\text{work done (J)}}{\text{time taken (s)}}$$

(b) (i) The extra power is proportional to the speed. If the speed doubles, the extra power needed doubles.
   (ii) No. If the speed doubles, the power needed to overcome resistance does not double.

(c) (i) Yes. To climb it at 2.2 m/s on an ordinary bicycle you need 215 W and on a lightweight bicycle you need only 193 W, so you could climb it at a lower speed.
   (ii) No. You need more than 220 W to climb it at 4.4 m/s, so you couldn't climb it at 5 m/s.

(d)

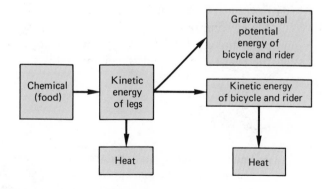

(e) The forces due to air resistance opposing your motion are less.
(f) If the pedals rotate twice, the chain passes over 100 teeth. This is five revolutions of the rear gear and the rear wheel therefore rotates five times.

47

Example 4.11

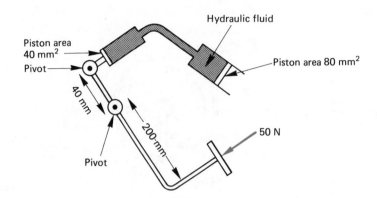

The diagram represents the principle of a car hydraulic braking system. A force of 50 N is applied to the brake pedal, which is at one end of a lever arm. The other end of the lever arm is connected to a piston of area 40 mm^2. A hydraulic fluid fills the space between this piston and another one of area 80 mm^2. Using the measurements shown in the diagram and ignoring any frictional forces,

(a) State the principle of moments and use it to calculate the force on the piston of area 40 mm^2. **(4 marks)**

(b) What is the pressure on the piston of area 40 mm^2? **(2 marks)**

(c) What is the pressure on the piston of area 80 mm^2? **(2 marks)**

(d) What is the force on the piston of area 80 mm^2? **(2 marks)**

(e) Why would it be dangerous if some air got trapped in the hydraulic fluid? **(2 marks)**

Solution 4.11

(a) When a body is in equilibrium,

anticlockwise moments = clockwise moments

50 N × 200 mm = force × 40 mm

$$\text{force} = \frac{50 \text{ N} \times 200 \text{ mm}}{40 \text{ mm}} = 250 \text{ N}$$

(b) Pressure = $\dfrac{\text{force}}{\text{area}}$ = $\dfrac{250 \text{ N}}{40 \text{ mm}^2}$ = 6.25 N/mm^2

(c) 6.25 N/mm^2 [The pressure is transmitted throughout the fluid: see Section 2.4.]

(d) Force = 6.25 N/mm^2 × 80 mm^2 = 500 N

(e) Air is compressible and if the brake pedal was pushed the force would compress the air. Little braking force would be applied to the second piston unless the brake pedal could travel a very long way.

Example 4.12

(a) Describe an experiment you would do in the laboratory to determine the position of the centre of mass of a semicircular sheet of plastic (e.g. a mathematical protractor).

(b) A painter stands on a horizontal platform which has a mass of 20 kg and is 5.0 m long. The platform is suspended by two vertical ropes, one attached to each end of the platform. The mass of the painter is 70 kg. If he is standing 2.0 m from the centre of the platform, calculate the tension in each of the ropes.

Solution 4.12

(a) Make holes at three points near the perimeter and well spaced out round it. Suspend the sheet by a fixed horizontal pin passing through one of the holes. Make sure the sheet can swing freely in a vertical plane. Hang a plumb line from the pin, and when the sheet is stationary, mark on it the position of the plumb line. Repeat this procedure for each of the other holes in turn. The centre of mass is where the three lines showing the positions of the plumb line cross.

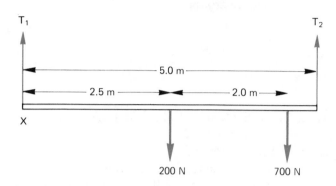

(b) The diagram shows the forces acting on the platform. The platform is in equilibrium, so, taking moments about X,

anticlockwise moments = clockwise moments
$T_2 \times 5$ m $= (200 \times 2.5)$ N m $+ (700 \times 4.5)$ N m
5 m $\times T_2 = 500$ N m $+ 3150$ N m
$T_2 = \dfrac{3650 \text{ N m}}{5 \text{ m}}$
$T_2 = 730$ N

Since the platform is in equilibrium, the total force upwards must equal the total force downwards; hence,

200 N $+ 700$ N $= 730$ N $+ T_1$
$T_1 = 170$ N

The tensions are 730 N in the rope nearest the painter, and 170 N in the rope furthest from the painter.

[There are two unknowns in this problem, namely T_1 and T_2. To obtain an equation with only one unknown, choose a point about which to take moments so that the moment of one of the forces is zero. In this case the moment of T_1 about X is zero. T_2 could have been calculated by taking moments about the other end of the platform.]

4.6 Have You Mastered the Basics?

1. Can you explain the meaning of work, power, energy, joule, watt, principle of moments, centre of gravity, and efficiency?
2. Can you discuss the energy changes which take place at a power station, and the world's main sources of energy?
3. Do you know the difference between kinetic energy and potential energy, and do you understand the formula by which they may be calculated?
4. A uniform plank is balanced at its midpoint. Jack, who weighs 500 N, sits 2 m from the centre and Jill, who weighs 400 N, sits on the same side 3 m from the

centre. Where must John, who weighs 550 N, sit if the plank is balanced horizontally?

5. A force of 6 N moves through 3 m in 2 s. What is (i) the work done, (ii) the power developed?

6. A mass of 1.2 kg is raised 4 m by an effort of 3 N moving through 20 m. What is (i) the potential energy gained by the load, (ii) the work done by the effort, (iii) the efficiency of the system?

7. A mass of 2 kg is lifted vertically 3 m and then allowed to fall. What is (i) the potential energy when it is 3 m above the ground, (ii) the velocity just before it hits the ground?

4.7 Answers and Hints on Solutions to 'Have You Mastered the Basics?'

1. See Sections 4.1, 4.2 and 4.4.
2. See Section 4.2.
3. See Section 4.2.
4. Take moments about the centre. Clockwise moments = Anticlockwise moments. (500×2) N m + (400×3) N m = $x \times 550$ N, where x is John's distance from the centre. Answer = 4 m from the centre on the other side of the plank to Jack and Jill.
5. (i) Work done = force × distance = 18 J. (ii) Substitute in the equation for power in Section 4.2. Power = 18 J/2 s = 9 W.
6. (i) Potential energy = mgh = 1.2 kg × 10 N/kg × 4 m = 48 J. (ii) Use work done = force × distance, to calculate the work done by the effort. Answer = 60 J. (iii) Use the equation in Section 4.4 to calculate the efficiency. Answer = 0.8 or 80%.
7. (i) Potential energy = mgh (see Section 4.2) = $(2 \times 10 \times 3)$ J = 60 J. (ii) Kinetic energy = $\frac{1}{2}mv^2$ [see Section 4.2] and this equals 60 J. Hence, $v = \sqrt{60}$ m/s = 7.7 m/s.

4.8 Questions

(Answers and hints on solutions will be found in Section 4.9. Where necessary, take the Earth's gravitational field strength as 10 N/kg.)

Question 4.1

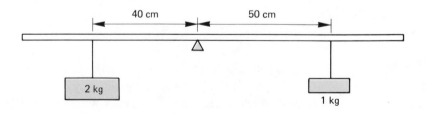

The diagram shows a uniform beam which is pivoted at its mid-point and from which two loads are suspended. If a third load is attached to balance the system, its mass and distance from the mid-point may be

A 1 kg, 10 cm   B 1 kg, 50 cm   C 2 kg, 10 cm   D 2 kg, 30 cm   E 3 kg, 10 cm   (AEB)

50

Question 4.2

A mass of 2 kg is 5 m above the floor and falling vertically with a velocity of 20 m/s. What is (i) its kinetic energy and (ii) its potential energy? **(6 marks)**

Question 4.3

An object is acted on by a retarding force of 10 N and at a particular instant its kinetic energy is 6 J. The object will come to rest after it has travelled a further distance, in m, of

A $\frac{3}{5}$ B $1\frac{2}{3}$ C 4   D 16   E 60   (AEB)

Question 4.4

In an experiment with a certain machine it was found that an effort of 10 N just moved a load of 40 N and that the load was moved 1 m when the effort moved 10 m. The efficiency of this machine is

A 0.4%   B 4%   C 10%   D 25%   E 40%   (AEB)

Question 4.5

(a) An Inter-City train goes up an incline which raises its vertical height 12 m. If the weight of the train is 5×10^6 N, how much work is done against the force of gravity? **(3 marks)**

(b) If the actual distance travelled up the incline is 12 000 m, and the train travels at 60 m/s, how long will it take to get to the top of the incline? **(2 marks)**

(c) What is the minimum power that the engine must produce in order to climb up the incline? **(3 marks)**

(d) Why is the word 'minimum' important in (c) above? **(2 marks)**

Question 4.6

(a)

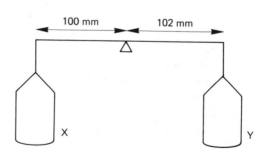

The diagram shows a badly manufactured beam balance with unequal arms. The beam balances with no masses in the pans X and Y. Suggest one reason why this is possible. If a mass of 300 g is placed in pan Y, calculate the mass which must be placed in pan X to restore the balance. **(5 marks)**

51

(b) An object put on a beam balance gives the same reading no matter where it is on the Earth's surface, but if it is on a very sensitive spring balance, the reading varies from place to place. Explain why this is so. **(5 marks)**

(c) Suppose that an object is attached to a spring balance on the Earth's surface and the Earth's speed of rotation is gradually increased. What effect, if any, will this have on
   (i) the gravitational attraction between the object and the Earth, and
   (ii) the reading of the spring balance?
   (Give reasons for your answers.) **(6 marks)**

(d) A person of mass 60 kg stands on a spring weighing machine inside a lift which is accelerating upwards at 3 m/s². Calculate the reading (in newtons) of the weighing machine. **(4 marks)**

(L)

Question 4.7

The two main forces opposing the motion of a bicycle are wind resistance and rolling resistance (rolling resistance is caused by the tyres rolling along the road). The table below gives the value of these forces, and the power required to propel the bicycle against these forces for two typical bicycles. The third set of data is for a bicycle that has somehow found its way onto a surfaced road on the moon.

The data have been calculated for four different speeds, and the speeds are given at the top of the columns.

Bicycles with upright handlebars have fairly large soft tyres. Racing bicycles have much narrower tyres, which are inflated to a high pressure.

10 m/s is about 22 miles/h. 750 W is 1 h.p.

	Mass/kg	Speed	2.5 m/s	5.0 m/s	7.5 m/s	10.0 m/s
Upright bicycle	18	wind force/N	2.1	8.4	19.2	34
rider	73	rolling force/N	5.3	5.3	5.3	5.3
		power/W	18.5	68.5	184	393
Racing bicycle	9	wind force/N	1.2	4.8	10.9	19
rider	73	rolling force/N	2.4	2.4	2.4	2.4
		power/W	9.0	36	100	214
Moon bicycle	9	wind force/N	0.00	0.00	0.00	0.00
rider	73	rolling force/N	0.67	0.67	0.67	0.67
spacesuit	7	power/W				

(a) Explain why the rolling resistance for the upright bicycle is greater than it is for the racing bicycle. **(2 marks)**

(b) Why is the wind resistance for the upright bicycle much greater than it is for the racing bicycle? **(1 mark)**

(c) Why is the wind resistance on the Moon zero? **(1 mark)**

(d) Explain why the rolling resistance for the bicycle on the Moon is much less than the rolling resistance on the Earth. **(2 marks)**

(e) How does the rolling resistance vary with the speed? **(1 mark)**

(f) How does the wind resistance vary with the speed? **(2 marks)**

(g) Suppose that your friend was planning a bicycle ride on an upright bicycle. How long would it take him to cycle 36 km at 10 m/s? Explain, using the data in the table, why he would have to cycle more slowly than this. **(3 marks)**

(h) What forward force must result from the force he applied to the pedals if he is to cycle at a steady speed of 5.0 m/s on a racing bicycle? **(2 marks)**

(i) If the net forward force on a racing bicycle is 3.0 N, what would the acceleration be? **(3 marks)**

(j) The last line of information about the Moon bicycle has been left blank. Fill in the power for each of the speeds. **(3 marks)**

4.9 Answers and Hints on Solutions to Questions

1. Difference between the two turning effects is (40 x 2) kg cm − (50 x 1) kg cm = 30 kg cm.
   Answer **E**

2. See Section 4.2. Kinetic energy = 400 J, potential energy = 100 J.

3. Work done = force x distance [see Section 4.2]; 6 J = 10 N x distance; hence, distance = 3/5 m.
   Answer **A**

4. Work got out of machine = energy gained by load = 40 N x 1 m = 40 J.
   Work put into machine = work done by effort = 10 N x 10 m = 100 J.
   Efficiency = 40/100 = 0.4 = 40%.
   Answer **E**

5. (a) A force of 5×10^6 N is moved through 12 m.

   Work = force x distance = 5×10^6 N x 12 m = 6×10^7 J.

   (b) It takes the train 1 s to go 60 m, so every 60 m takes 1 s.

   $$\text{Time} = \frac{12\,000\ \text{m}}{60\ \text{m/s}} = 200\ \text{s}$$

   (c) $\text{Power} = \dfrac{6 \times 10^7\ \text{J}}{200\ \text{s}} = 300\,000$ W

   (d) In practice more power would be needed, because work has to be done in overcoming the frictional forces.

6. (a) The balance pan X has a greater mass than balance pan Y, or the beam is not uniform. 306 g.

   (b) The extension of a spring balance depends on the gravitational force acting on the mass hung from its end and this force varies over the Earth's surface. With a beam balance the force on the object and on the equal mass on the other pan is always the same and the beam remains balanced.

   (c) (i) The gravitational attraction will not change, since gravitational attraction depends only on the mass and its distance from the centre of the Earth. (ii) The spring balance would read less. With increasing speed of rotation of the Earth, more and more of the Earth's gravitational force is needed to provide the force required (centripetal force) to keep the object moving in a circle. The object's weight, which is the difference between the total gravitational force and the centripetal force, is less, and the spring balance therefore reads less.

   (d) Apply $F = ma$ [Section 3.2], where F is the resultant force on the person (the force upwards of the scales minus his weight). Let R be the reading (the push upwards of the scales) on the weighing machine; then $R − 600$ N = 60 kg x 3 m/s^2; hence, $R = 780$ N.

7. (a) A greater area of tyre in contact with the road.

   (b) A greater effective frontal area.

   (c) There is no air on the Moon.

   (d) The bicycle weighs less on the Moon, so the tyres would distort less.

   (e) There is no variation.

   (f) A square law; if the speed is doubled, the wind resistance goes up by a factor of 4.

   (g) 36 000 m ÷ 10 m/s = 3600 s = 1 h. You can't produce more than half a horsepower continuously for 1 h.

(h) For a constant velocity the resultant force must be zero. 7.2 N.

(i) $F = ma \Rightarrow a = 0.037 \text{ m/s}^2$.

(j) Use power = force × velocity; see Section 4.2. 1.7 W, 5.0 W and 6.7 W.

5 Expansion, Gas Laws, Thermometers, Kinetic Theory and Molecular Size

5.1 Expansion

Gases, solids and liquids usually expand when heated. For a given temperature rise, gases expand more than liquids and liquids expand more than solids. Water has the unusual property of contracting as it is heated from 0°C to 4°C; it has its maximum density at 4°C.

A bimetallic strip may be used to construct a thermostat (see Example 7.6). The bimetallic strip curves as the temperature changes, and this movement is used to make and break a contact.

5.2 The Behaviour of Gases

When the temperature, T, of a gas is raised, either the pressure, p, or the volume, V, or both increase. They are related by the equation

$$\frac{p_1 V_1}{T_1} = \frac{p_2 V_2}{T_2}$$

where the suffix 1 represents the initial state of the gas and the suffix 2 represents the final state. *Remember that T must be in kelvin — that is (°C + 273).*

For a fixed mass of gas at constant temperature, $p_1 V_1 = p_2 V_2$, or $p \propto \dfrac{1}{V}$

(Boyle's law). For a fixed mass of gas at constant volume, $p \propto T$ (the pressure law). For a fixed mass of gas at constant pressure, $V \propto T$ (Charles's law).

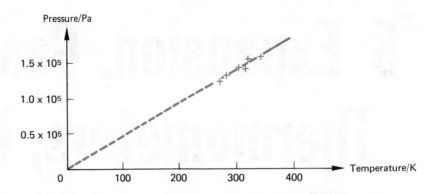

Figure 5.1 Graph of pressure against temperature for a fixed mass of gas at constant volume. A graph of volume against temperature (pressure constant) is the same shape and goes through 0 K.

Figure 5.1 is a graph of pressure (p) against temperature (T), the volume being kept constant. The dotted part of the line is what we might expect to happen if we went on cooling the gas and it never liquefied. The absolute zero of temperature is the temperature at which the pressure would become zero.

5.3 Thermometers

Common types of thermometer are (a) mercury or alcohol in glass (both depend for their working on the fact that liquids expand when heated); (b) thermistors (they depend for their working on the change of resistance of semiconductors with temperature); (c) thermocouples (they depend for their working on a voltage that develops across junctions of two different metals as the temperature changes).

In order to calibrate a thermometer, two fixed points are needed.

The *lower fixed point* is the temperature at which pure ice melts at normal atmospheric pressure.

The *upper fixed point* is the temperature of steam from water boiling at normal atmospheric pressure.

To calibrate a thermometer in °C, mark the lower fixed point (0°C) and the upper fixed point (100°C) and then mark one hundred equal divisions between the two points.

5.4 Kinetic Theory

The molecules of a solid vibrate about fixed positions. When heat is supplied to a solid, the molecules vibrate faster and through greater distances than before. The molecules are pushed further apart and the solid expands. If sufficient heat energy is supplied, the solid melts. In a liquid the molecules no longer vibrate about fixed positions and move freely among one another. The forces between the molecules are enough to give a liquid a definite volume but not a definite shape. As more heat energy is supplied, the molecules move faster and gain enough energy to break right away from one another and move independently. The liquid has started to boil. The molecules of a gas are in random motion, with varying velocities. They are continually colliding with one another and with the walls of the container, causing a pressure on the walls of the containing vessel. If the temperature is increased, the molecules move faster and hit the walls more often and harder, causing the pressure to increase. At the absolute zero of temperature (0 K or −273°C) the molecules have their lowest possible kinetic energy.

Evaporation is the escape of the faster-moving molecules from the surface of a liquid, and this takes place at all temperatures. The average kinetic energy of the remaining molecules has decreased and so the temperature has fallen. The rate of evaporation may be increased by (i) increasing the surface area of the liquid, (ii) blowing air over the surface or (iii) increasing the temperature of the liquid. Boiling occurs when bubbles of vapour form in the body of the liquid, and rise and escape from the surface of the liquid. Boiling takes place at a particular temperature.

5.5 Evidence for Kinetic Theory

(a) Brownian Motion (see Examples 5.5 and 5.6)

This provides evidence for the kinetic theory of matter. Brownian motion may be observed by using a microscope to view illuminated smoke particles in a smoke cell. The specks of light (the light reflected and scattered by the smoke particles) jostle about in a random irregular manner as they are bombarded by the invisible moving air molecules.

(b) Diffusion (see Example 5.7)

This also provides evidence for the motion of molecules.

5.6 Molecular Size: The Oil Drop Experiment (see Example 5.7)

An oil drop of measured radius is dropped into the middle of a large area of clean water which has a light powder sprinkled on its surface. By using the equation

volume of drop = volume of oil on surface = (area of oil on surface) × thickness

the thickness of the film may be calculated. The volume of the drop is calculated from its diameter, and the area of the film on the surface from the diameter of the circle of oil on the surface. The thickness of the film is an indication of the size of a molecule.

5.7 Worked Examples

Example 5.1

Describe how you would investigate the relationship between the volume and pressure of a fixed mass of dry air at a constant temperature. Show clearly how the values of volume and pressure are measured and state any precautions which are taken.   (8 marks)
Assuming that you have been provided with the following results, use them to obtain a *straight line* graph. What two features of the graph show that Boyle's law holds?

$Pressure/10^4$ Pa	2.0	3.0	4.0	5.0	6.0	8.0
$Volume/10^{-4}$ m³	4.0	2.7	2.0	1.6	1.3	1.0

(8 marks)
(L, part question)

The diagram shows the apparatus. The volume of the trapped air is read on the graduated scale behind the air column and the pressure is read on the Bourdon gauge. The tap is opened and a car pump is used to force the oil up the tube, decreasing the volume of the gas. The volume and pressure are again read. By opening and closing the tap a series of readings are obtained. Each

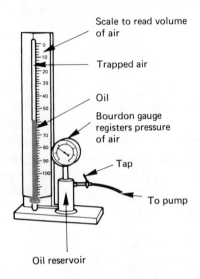

time the oil is lowered in the tube a short time must elapse before the reading of the volume is taken. This allows time for the oil on the sides of the tube to run down the tube and time for the air inside the tube to reach room temperature. Throughout the experiment it is important to check that the temperature of the laboratory remains constant.

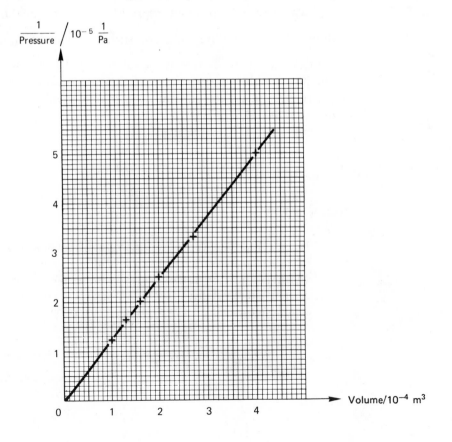

$\dfrac{1}{Pressure}$ $\Big/ 10^{-5}\ \dfrac{1}{Pa}$	5.0	3.3	2.5	2.0	1.6(6)	1.2(5)
$Volume/10^{-4}\ m^3$	4.0	2.7	2.0	1.6	1.3	1.0

The graph is a straight line through the origin, confirming that the volume is inversely proportional to the pressure.

Example 5.2

Which of the graphs below best represents the results of an experiment to verify Boyle's law?

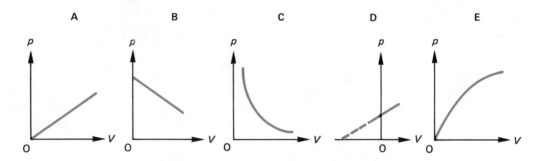

Solution 5.2

[A graph like the one in C is always obtained when two variables are inversely proportional.]
Answer **C**

Example 5.3

Which one of the following graphs correctly represents the variation of pressure with absolute temperature for a fixed mass of an ideal gas at constant volume?

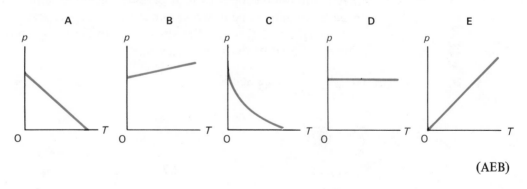

(AEB)

Solution 5.3

[The pressure is directly proportional to the kelvin (absolute) temperature. Direct proportion is represented by a straight line graph through the origin.]

Answer **E**

Example 5.4

(a) A pupil investigates the relationship between the pressure and temperature of a fixed mass of gas, using the apparatus shown.

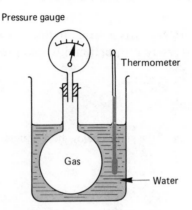

Pressure gauge

Thermometer

Gas

Water

He heats the water continuously with a Bunsen burner and records the pressure and temperature readings every minute.
  (i) State two ways in which this experiment may be improved.
  (ii) Using the results from an improved experiment, describe how the relationship between pressure and temperature on the kelvin scale may be found.   **(4 marks)**

(b) A cylinder of oxygen at 27°C has a gas pressure of 3×10^6 Pa.
  (i) Calculate the pressure of the gas if the cylinder is cooled to 0°C.
  (ii) Describe what happens to the gas molecules as the gas is cooled and indicate how this results in a reduction of pressure.   **(4 marks)**

(c) When air in a bicycle pump is compressed by moving the piston, the temperature of the air in the pump increases. Explain this temperature rise in terms of the kinetic theory of gases.   **(2 marks)**

(SEB)

Solution 5.4

(a) (i) Include a stirrer to ensure that all the water is at the same temperature. Remove the bunsen burner and stir for some time before taking a reading.
  (ii) A graph should be drawn of pressure against temperature, with the temperature axis going down to about −300°C. The graph will be found to be a straight line through the absolute zero. This shows that the pressure is proportional to the kelvin temperature.

(b) (i) $\dfrac{p_1}{T_1} = \dfrac{p_2}{T_2}$ [see Section 5.2]

$$\therefore \frac{3 \times 10^6 \text{ Pa}}{(273 + 27) \text{ K}} = \frac{p_2}{273 \text{ K}}$$

$$\therefore p_2 = \frac{273 \times 3 \times 10^6}{(273 + 27)} \text{ Pa} = 2.7 \times 10^6 \text{ Pa}$$

  (ii) As the gas is cooled the average kinetic energy of the gas molecules decreases, so the average velocity of the molecules falls. The impacts on the side of the vessel are not so hard and are less frequent, so the pressure falls.

(c) As the moving piston strikes the moving molecules, the velocity of the molecules is increased. The average kinetic energy of the molecules has increased, and therefore the temperature has increased.

Example 5.5

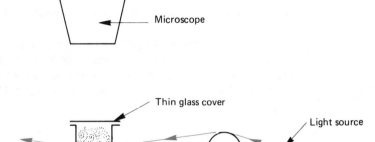

Microscope

Thin glass cover

Light source

Small cell containing
air and smoke

Glass rod

The diagram shows one form of apparatus used to observe the Brownian motion of smoke particles in air. A student looking through the microscope sees tiny bright specks which he describes as 'dancing about'.
(a) What are the bright specks?
(b) Why are the specks 'dancing about'?
(c) What is the purpose of the thin glass cover? **(5 marks)**
(L)

Solution 5.5

(a) The specks of light are light which is reflected and scattered by the smoke particles.
(b) The smoke particles are being bombarded by the air molecules which are in rapid random motion.
(c) To reduce air (convection) currents and to stop the smoke escaping.

Example 5.6

Brownian motion is observed by observing smoke particles under a microscope. Which of the following statements is correct?
A The smoke particles move about with uniform velocity
B The motion is caused by the air molecules colliding with the smoke particles
C If larger smoke particles were used, the Brownian motion would become more rapid
D The experiment would work just as well in a vacuum

Solution 5.6

A is incorrect and B is correct (see Section 5.5). Larger smoke particles would show *less* rapid motion. They are more massive than smaller particles and are therefore accelerated less for a given force (see Section 3.2). In a vacuum there would be no air particles to bombard the smoke particles.
Answer **B**

Example 5.7

(a) Describe one experiment to demonstrate the diffusion of gases and one experiment to demonstrate the diffusion of liquids. Each answer must include a diagram of the apparatus. **(8 marks)**

(b) How is diffusion explained by the kinetic theory of gases? **(4 marks)**

(c) In the oil drop experiment to estimate the length of an oil molecule a drop of oil of volume 0.06 mm^3 is dropped on to the surface of water. A patch of oil of area 45 000 mm^2 is formed on the surface of the water. Use these figures to estimate the length of an oil molecule. **(8 marks)**

Solution 5.7

(a) ***Diffusion of Gases***

A capsule of bromine is placed in the rubber tubing and the end of the tubing sealed with a glass rod. The tap is opened and the bromine capsule broken by squeezing the rubber tube with a pair of pliers. The liquid bromine escapes into the glass tube. The dark colour of the bromine vapour will be seen to move slowly (diffuse) up the tube.

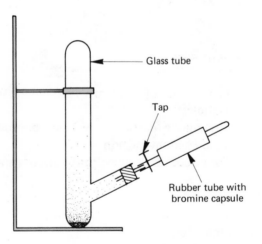

Diffusion of Liquids

A strong sugar solution is poured into a gas jar. A strong copper sulphate solution is poured very slowly and very carefully on top of the sugar solution. Water is very carefully poured on top of the copper sulphate solution. If these operations are carried out carefully there will be a fairly sharp dividing line between the layers. The blue copper sulphate will be observed to diffuse downwards and upwards.

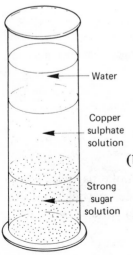

(b) When two substances whose molecules are in motion are put in contact with each other, the moving molecules pass across the boundary between the two substances. Some will collide almost immediately with molecules of the other substance, others will move a little further in the spaces between the molecules before colliding with another molecule. In this way the molecules of one substance gradually move into the other substance.

Let thickness of oil on surface = x mm
But volume of oil on surface = volume of oil drop [see Section 5.6]
∴ 0.06 mm³ = 45 000 mm² × x

$$\therefore x = \frac{0.06 \text{ mm}^3}{45\,000 \text{ mm}^2} = 1.3 \times 10^{-6} \text{ mm}$$

Length of oil molecule is about 1.3×10^{-6} mm

Example 5.8

Use the kinetic theory to explain
(a) why air kept at constant temperature and constant volume exerts a constant pressure;   (4 marks)
(b) why heat energy must be supplied to turn water into steam;   (4 marks)
(c) the existence of an absolute zero of temperature.   (4 marks)

Solution 5.8

(a) The molecules of air are in rapid motion with a large range of speeds. Large numbers strike the walls every second. The average speed of the molecules is constant, so the average force due to the bombardment is constant. The pressure is the force per unit area, so the pressure is constant.
(b) Intermolecular forces maintain the shape of the liquid. When heat energy is supplied, the molecules gain kinetic energy and eventually break away from the attractive forces which exist between them when they are in the liquid state. They break free, and so the water has become steam. Energy is also used to push back the air molecules above the liquid.
(c) As heat is withdrawn from a substance, the energy of the molecules becomes less and their velocity decreases. Eventually there comes a time when no more energy can be withdrawn and we have reached the absolute zero of temperature.

5.8 Have You Mastered the Basics?

1. Do you understand the gas laws?
2. Can you distinguish between evaporation and boiling?
3. Can you describe the oil drop experiment, experiments to demonstrate diffusion and Brownian motion?
4. Can you use the kinetic theory to explain diffusion, Brownian motion, evaporation, boiling and the existence of absolute zero?
5. Can you use the kinetic theory of gases to explain how the pressure of a gas in a container changes if the temperature is changed?
6. A fixed mass of gas at 0°C has a volume of 546 cm³ at a pressure of 3 atmospheres. It is heated until its temperature reaches 77°C. If the new volume is 1050 cm³, what is the new pressure?

5.9 Answers and Hints on Solutions to 'Have You Mastered the Basics?'

1. See Section 5.2 and Examples 5.1, 5.2, 5.3 and 5.4.
2. See Section 5.4.

3. See Sections 5.5 and 5.6 and Examples 5.5, 5.6 and 5.7.
4. See Sections 5.4 and 5.5.
5. See Section 5.4.
6. Substitute in the equation in Section 5.2. Don't forget to put the temperatures in kelvin. T_1 is $273°C$ and T_2 is $(273 + 77)°C$.
   Answer   2 atmospheres

5.10 Questions

(Answers and hints on solutions will be found in Section 5.11.)

Question 5.1

(a) Describe how you would investigate the variation of volume with pressure of a fixed mass of air kept at constant temperature.
(b) The table below shows a series of readings of pressure and volume for such an experiment. Make a table of pressure and $\dfrac{1}{\text{volume}}$. Plot a graph of pressure against $\dfrac{1}{\text{volume}}$ and use your graph to determine the volume when the pressure is 160 kN/m^2.

$Pressure \Big/ \dfrac{\text{kN}}{\text{m}^2}$	400	267	200	120
$Volume/\text{mm}^3$	3.0	4.5	6.0	10.0

Question 5.2

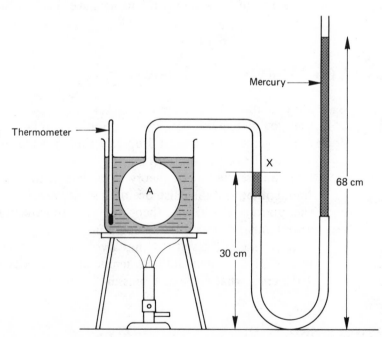

(a) The diagram shows an apparatus which can be used to investigate how the pressure of a fixed mass of air at a constant volume trapped in the bulb A varies with temperature.
   (i) How would you ensure that all the air trapped in the bulb A was at the temperature recorded by the thermometer?   **(2 marks)**

(ii) Why, when taking readings of the pressure, is it necessary to ensure that the mercury level coincides each time with the mark X? **(2 marks)**

(iii) Using the values given in the diagram, calculate the total pressure of the air trapped in the bulb **A**, given that the height of a mercury barometer in the laboratory is 76 cm. (The pressure of 76 cm of mercury is equivalent to 1×10^5 Pa.) **(3 marks)**

(iv) Sketch a graph of the results that would be obtained from such an experiment. What conclusion could be drawn from the graph? What do you understand by absolute zero of temperature? **(6 marks)**

(v) Explain briefly how the kinetic theory of gases accounts for the results obtained in the above experiment. **(4 marks)**

(b) A bottle is corked when the air inside is at $20°C$ and the pressure is 1.0×10^5 Pa. If the cork blows out when the pressure is 3.0×10^5 Pa, calculate the temperature to which the bottle must be heated for this to happen. (Assume that the bottle does not expand.) **(3 marks)**

(L)

Question 5.3

PART I

(a) The diagram shows a bimetal strip wound into a flat spiral. Metal **A** has a higher expansivity than metal **B**.

Describe and explain what happens when the strip is heated. **(3 marks)**

(b) When constructing a thermometer for normal laboratory use, the bulb is made of thin glass, the bore of the tube is narrow, and mercury is used as the thermometric liquid. Explain why

(i) the bulb is of thin glass,

(ii) the bore is narrow,

(iii) mercury is chosen. **(5 marks)**

(AEB, part question)

Question 5.4

(a) (i) Sketch a graph to show how the pressure of a fixed mass of gas varies with the temperature as the temperature changes from $0°C$ to $100°C$, if the volume of the gas remains constant.

(ii) Sketch a graph to show how the volume of a fixed mass of gas varies with the pressure provided the temperature is kept constant. **(5 marks)**

(b) Explain, using the kinetic theory of gases, why the pressure of a fixed mass of gas increases as its temperature rises. **(4 marks)**

Question 5.5

(a) Describe a laboratory experiment which demonstrates Brownian motion. State clearly what you observe and explain your observations.

(b) Describe an experiment to illustrate diffusion in a gas.

(c) A cylinder contains 60 litres of air when the pressure is 2×10^5 Pa and the temperature $27°C$. What will the temperature in the cylinder be when the volume is 180 litres and the pressure is 10^5 Pa?

5.11 Answers and Hints on Solutions to Questions

1. See Example 5.1. 7.5 mm^3.

2. (a) (i) Remove the bunsen and stir for some time before taking a reading. (ii) This keeps the volume constant. (iii) The pressure of the trapped air is (atmospheric + pressure due to 38 cmHg) = 76 + 38 cmHg = 1.5×10^5 Pa. (iv) See Section 5.2. (v) See Section 5.4.

   (b) Use $p_1/T_1 = p_2/T_2$; temperature is 606°C.

3. (a) Because A expands more than B, the curvature of the metal will decrease and the strip will uncurl.

   (b) (i) Heat will conduct through the thin glass quicker than if the glass were thick and the thermometer will reach the temperature it is measuring quickly. (ii) The mercury moves further for a given change in temperature and hence the markings on the scale are further apart and the thermometer is more sensitive. (iii) Mercury has a high boiling point and a reasonably low freezing point; a mercury thermometer therefore has a large range. Mercury is easily visible and doesn't stick to the sides of the tube.

4. (a) (i) See Section 5.2. (ii) See Example 5.2.

   (b) See Section 5.4.

5. (a) See Section 5.5 and Examples 5.5 and 5.6.

   (b) See Example 5.7.

   (c) Use pV/T is constant. Temperature = 450 K = 177°C.

6 Specific Heat Capacity, Specific Latent Heat, Refrigeration, Boiling and Melting

6.1 Heat and Temperature

If a large saucepan full of water is heated and a small saucepan full of water is supplied with the same amount of heat energy, then the temperature of the water in the smaller saucepan rises more than the temperature of the water in the large saucepan. Heat is a form of energy (unit, J) and temperature is the degree of hotness (unit, °C or K).

6.2 Specific Heat Capacity

The *specific heat capacity* of a substance is the heat energy required to raise 1 kg of it through 1 K. Its units are J/(kg K).

It follows that if a mass m of a substance of specific heat capacity c is raised through a temperature T, then

heat required, $Q = mcT$

To measure c, heat may be applied to a known mass, m, of substance by an immersion heater connected to the power supply via a joulemeter (which measures the energy supplied, Q). The rise in temperature, T, is measured, and the equation above is used to calculate c (see Example 6.3). If the substance is a solid, a hole is drilled in a block of it so that the immersion heater may be inserted. The thermometer is inserted into a second hole in the block.

Water has a high specific heat capacity compared with other liquids. Hence, for a given mass and a given quantity of heat, the temperature rise is comparatively small. Climate is therefore more temperate near large expanses of water, and fruit growing districts are often near large lakes.

Heating and cooling systems which use liquids work better the greater the mass of liquid used and the greater the specific heat capacity of the liquid.

6.3 Specific Latent Heat

The *specific latent heat* is the quantity of heat required to change the state of 1 kg of substance without change in temperature. Its units are J/kg (you might meet kJ or MJ; 1 kJ = 10^3 J and 1 MJ = 10^6 J). If the change is from liquid to vapour, then it is the specific latent heat of vaporisation (note the spelling!), and if the change is from solid to liquid, then it is the specific latent heat of fusion. It follows that to change the state of a mass m of a substance of specific latent heat L

heat required, $Q = mL$

The specific latent heat of vaporisation of water is a large value (over 2 MJ), and much more heat is needed to boil away 1 kg of water than to raise its temperature 100°C.

The measurement of the specific latent heat of vaporisation is described in Example 6.6. The latent heat of fusion of ice may be measured by putting an immersion heater in a funnel and surrounding it with ice. The water from the melted ice is collected and weighed. The heat supplied (measured on a joulemeter) to melt a known mass of ice (m kg) is known and hence L can be calculated from the equation $Q = mL$. (In practice it is best to have a similar funnel packed with ice without a heater, so that the mass of ice which melts in the absence of a heater may be subtracted from m, in order to get the mass of ice melted by the heater.)

6.4 Effect of Impurities and Pressure on the Boiling and Melting Points

If salt is added to water it raises the boiling point and lowers the freezing point. Salted water for cooking vegetables therefore boils above 100°C and salt put on roads in cold weather lowers the freezing point of water, and any ice or snow on the roads melts. Increase in pressure raises the boiling point and lowers the freezing point. In pressure cookers the water boils at a temperature above 100°C. Snowballs do not 'bind' on a very cold day because the increase in pressure on squeezing does not lower the freezing point below that of the snow, so the snow does not melt. Usually on squeezing, the snow melts and refreezes when the pressure is removed, so the snowball 'binds'.

6.5 Refrigerators

A compressor compresses a very volatile vapour (the refrigerant) and it liquefies. The refrigerant expands through a valve into tubes which are in the ice compartment. As it expands it again becomes a vapour. The energy needed to convert the liquid to a vapour is drawn from the ice compartment, which is therefore cooled. The vapour is compressed outside the refrigerator, where it gives out its latent heat (see Example 6.8 and Question 6.7).

6.6 Worked Examples

Example 6.1

5632 joules of heat energy raise the temperature of 0.4 kg of aluminium from 20°C to 36°C. The specific heat capacity of aluminium, in J/(kg K), is given by

$$\text{A } 5632 \times 0.4 \times 16 \qquad \text{B } \frac{5632}{0.4 \times 16} \qquad \text{C } \frac{5632 \times 16}{0.4} \qquad \text{D } \frac{5632 \times 0.4}{16} \qquad \text{E } \frac{0.4 \times 16}{5632}$$

(AEB)

Solution 6.1

$$\left[\begin{array}{l} Q = mcT \text{ (see Section 6.2)} \Rightarrow 5632 \text{ J} = 0.4 \text{ kg} \times c \times 16 \text{ K} \\ \qquad\qquad\qquad\qquad \Rightarrow c = \dfrac{5632}{0.4 \times 16} \text{ J/(kg K)} \end{array} \right]$$

<u>Answer **B**</u>

Example 6.2

A 50 W immersion heater is switched on and immersed in some water which is at 0°C. The graph below shows the temperature rise plotted against the time.

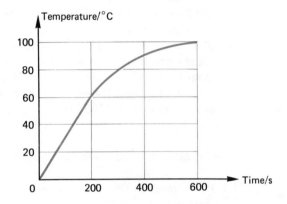

(i) How much energy is supplied by the heater in 400 s? **(3 marks)**

(ii) What is the rise in temperature during the first 200 s? **(1 mark)**

(iii) Complete the following table showing the energy supplied by the heater and the rise in temperature.

Time/s	0	200	400	600
Energy supplied by heater/kJ	0			
Temperature rise/K	0			

(5 marks)

(iv) Why is the temperature rise during the first 200 s greater than the rise in temperature during the last 200 s? **(3 marks)**

Solution 6.2

(i) [50 W is 50 J/s: see Section 4.2.]

Energy supplied = $\left(50 \dfrac{\text{J}}{\text{s}} \right) (400 \text{ s}) = 20\,000 \text{ J}$

(ii) 60°C.

Time/s	0	200	400	600
Energy supplied by heater/kJ	0	10	20	30
Temperature rise/K	0	60	90	100

(iv) More heat is escaping into the atmosphere because the body is at a higher temperature, so more heat must be supplied by the heater to obtain the same temperature rise. Also, the higher the temperature, the faster the water is evaporating and more heat is supplied to change the state of the water.

Example 6.3

(a) (i) A 25 W immersion heater was placed in 100 g of a liquid and the temperature of the liquid rose by 20°C in 4 min. Calculate the specific heat capacity of the liquid.

(3 marks)

(ii) Assuming the figures given above are accurate, why is the result only an approximate one? **(2 marks)**

(iii) State two precautions necessary in doing this experiment in order to obtain an accurate result. **(2 marks)**

(b) In a domestic oil-fired boiler central heating system, 0.50 kg of water flows through the boiler every second. The water enters the boiler at a temperature of 30°C and leaves at a temperature of 70°C, re-entering the boiler after flowing round the radiators at 30°C. 3.0×10^7 J of heat is given to the water by each kilogram of oil burnt. The specific heat capacity of water is 4200 J/(kg K). Use this information to calculate

(i) the energy absorbed by the water every second as it passes through the boiler,

(3 marks)

(ii) the mass of oil which would need to be burnt in order to provide this energy.

(2 marks)

(c) If the owner of the above central heating system wished to reduce his fuel bill, suggest *three* ways he could do it. **(3 marks)**

Solution 6.3

(a) (i) Heat supplied by immersion heater in 4 min = $\left(25 \dfrac{\text{J}}{\text{s}}\right) \times (4 \times 60)$ s
$$= 6000 \text{ J}$$
Assume that all the heat supplied by the heater goes into the liquid
$Q = mcT$ [see Section 6.2]
6000 J = 0.1 kg × c × 20 K
[Don't forget that the mass must be in kg, and an interval of 1°C is the same as an interval of 1 K.]
$$c = \frac{6000 \text{ J}}{0.1 \text{ kg} \times 20 \text{ K}} = 3000 \text{ J/(kg K)}$$
The specific heat capacity of the liquid is 3000 J/(kg K).

(ii) The liquid must be in a container, and we have neglected the heat taken up by the container and also the heat lost to the atmosphere.

(iii) Lag the container to reduce the heat lost to the atmosphere and stir the liquid before taking the temperature.

(b) (i) $Q = mcT = \left(0.50 \dfrac{\text{kg}}{\text{s}}\right)\left(4200 \dfrac{\text{J}}{\text{kg K}}\right)(40 \text{ K}) = 84\,000 \text{ J/s}$

(ii) Mass = $\dfrac{84\,000 \text{ J}}{3.0 \times 10^7 \text{ J/kg}} = 0.0028 \text{ kg} = 2.8 \text{ g}$

(c) He could insulate his loft with fibre glass, double glaze his windows and put draught-proof material at the bottom of his doors.

[Alternatively, you could mention filling his walls with insulating material.]

Example 6.4

If the specific latent heat of steam at 100°C is 2.26 x 10⁶ J/kg the heat, in J, required to evaporate 2 g of water at 100°C is

A 2.00×10^2 **B** 1.13×10^3 **C** 4.52×10^3 **D** 1.13×10^6 **E** 4.52×10^6 (AEB)

Solution 6.4

[Heat supplied = mL (see Section 6.3) = $(0.002 \text{ kg}) \left(2.26 \times 10^6 \; \dfrac{J}{kg}\right) = 4.52 \times 10^3$ J.]

<u>Answer C</u>

Example 6.5

The graph shows the change in temperature when heat is applied at 20 000 joule/minute to 1 kilogram of a substance.

The specific latent heat of fusion of the substance in joule/kilogram is

A 2000   **B** 10 000   **C** 20 000   **D** 40 000   **E** 80 000   (AEB)

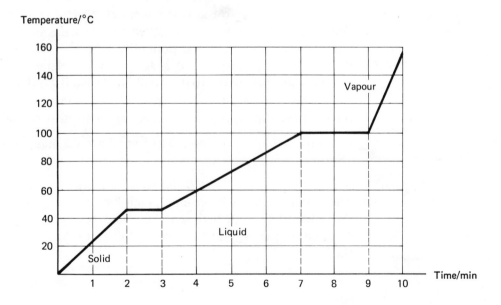

Solution 6.5

[It takes 1 min for the solid to melt and become a liquid. Heat supplied in 1 min = 20 000 J.]

<u>Answer C</u>

Example 6.6

An electric kettle about half filled with water was switched on and left until the water was boiling steadily. While the water was boiling steadily at 100°C, 250 g of water at 20°C was added to the water in the kettle. It took 30 s before the water was again boiling steadily.

(i) What is the power of the heating element in the kettle? [Specific heat capacity of water is 4200 J/(kg K).] **(6 marks)**

(ii) The power could have been found by finding the time for the water to rise from, say, 20°C to 40°C. What is the main advantage of the above method? **(3 marks)**

(iii) Describe an experiment using the same kettle which you could do to measure the specific latent heat of vaporisation of water at 100°C. **(7 marks)**

(iv) Explain in terms of the kinetic theory of gases why it is necessary to supply energy to change water to steam at 100°C. **(4 marks)**

Solution 6.6

(i) Heat supplied to water, $Q = mcT$ [see Section 6.2]
$$= (0.25 \times 4200 \times 80) \text{ J} = 84\,000 \text{ J}$$

[The water in the kettle at the time the cold water was poured in is at the same temperature at the beginning of the experiment as it is at the end of the experiment (100°C); hence, the heat supplied by the element is used to heat 0.25 kg of water through 80°C.]

$$\text{Power} = \text{rate of supply of heat} = \frac{\text{heat supplied}}{\text{time}} = \frac{84\,000 \text{ J}}{30 \text{ s}} = 2800 \text{ J/s}$$
$$= 2800 \text{ W}$$

(ii) If the time for the water to rise 20°C is measured, the kettle also takes up heat and this must be included in the calculation. In the above method the temperature of the kettle is the same at the end as it was at the beginning. During the experiment it has taken up the same amount of heat as it gave out when the cold water was poured in.

(iii) Weigh the kettle empty and again when the element is well covered with water. Bring the water to the boil and leave it boiling for, say, 3 minutes. Weigh the kettle again in order to determine the mass of water, m, which has boiled away. Then, if L is the specific latent heat of vaporisation of water,

heat supplied in 3 minutes = heat to boil away a mass of water
$2800 \times (3 \times 60) \text{ J} = m \times L$

m is known and therefore L can be calculated.

(iv) The molecules must be given enough kinetic energy (1) to break away from the attractive forces between them and become a vapour and (2) to give them enough energy to push back the atmosphere as they break away from the water.

Example 6.7

(a) Describe an experiment which shows that water can be made to boil at temperatures below 70°C. **(7 marks)**

(b) The table below shows the temperature of a liquid in a beaker as it cools in a laboratory.

Temperature/°C	86	60	55	55	55	55	49	41
Time/minutes	0	1	2	3	4	5	6	7

(i) Draw a graph of temperature against time.

(ii) Explain the shape of the graph. **(10 marks)**

Solution 6.7

(a) The apparatus is shown in the diagram. The flask initially contains water at 70°C. When the vacuum pump is switched on the pressure above the water is reduced. As the pressure above the water decreases the boiling point of the water goes down. The water begins to boil and continues boiling as the temperature and pressure further decrease.

(b) (i)

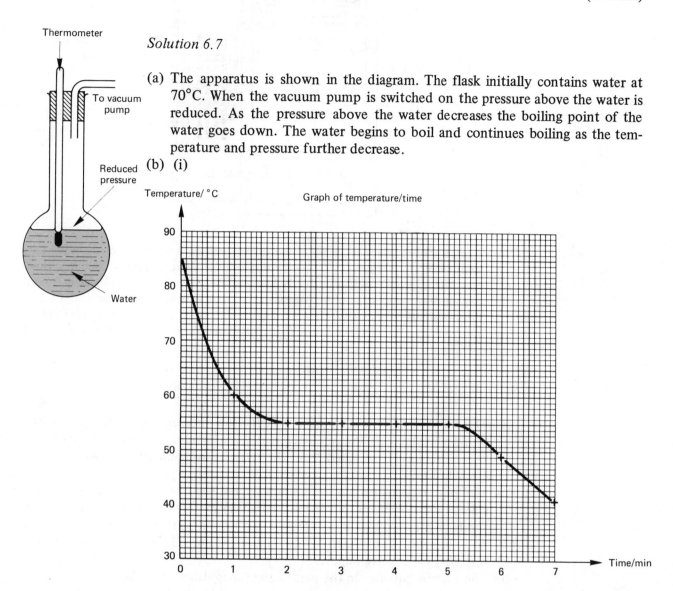

(ii) Heat is continually being lost from the beaker and passing into the atmosphere. The liquid loses heat and its temperature falls. When the freezing point is reached the temperature remains at 55°C until all the liquid has frozen. The latent heat of fusion is being released. Once the liquid has frozen the temperature of the solid falls as heat continues to pass into the atmosphere.

Example 6.8

The diagram shows the basic components of a refrigerator system which contains a volatile liquid known as the refrigerant.

(i) Explain how the action of the circulating refrigerant as it passes through the freezer unit reduces the temperature of its contents. **(4 marks)**

(ii) What is the purpose of the pump? **(2 marks)**

(iii) Explain why the external metal fins become warm while the refrigerator is in operation. What is the source of this heat energy? **(3 marks)**

73

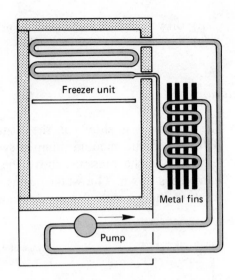

Freezer unit

Metal fins

Pump

(iv) When a tray of water is placed in the freezer unit the temperature of the water falls from room temperature (15°C) to freezing point in approximately 15 minutes. However, it is necessary to wait a further $1\frac{1}{2}$ hours before the ice becomes completely frozen. Explain why it is quite usual to wait such a long time for completely solid ice to form in the refrigerator. **(4 marks)**

(L, part question)

Solution 6.8

(i) The tubes in the freezer unit are on the low pressure side of the pump. The liquid entering this low pressure region evaporates and takes up its latent heat. This heat is supplied from the refrigerator compartment, which consequently cools.

(ii) The pump circulates the refrigerant. It reduces the pressure on the freezer side of the system so that the refrigerant evaporates, becoming a vapour. On the high pressure side it compresses the vapour so that it liquefies. [There is no expansion valve shown in the diagram. Compare the diagram in Question 6.7. Why is the valve important?]

(iii) The vapour liquefies in the part of the tubes attached to the fins. As it liquefies it gives out its latent heat. The metal tube and fins are good conductors of heat and are warmed up by the heat from the condensing refrigerant.

(iv) Heat is continually being withdrawn from the freezing compartment but because the latent heat of fusion of ice is large a lot of heat has to be withdrawn before all the water present has frozen. The heat that must be withdrawn to cool the water from 15°C to 0°C is about one-sixth of that needed to freeze the water at 0°C.

Ice is a poor conductor of heat, so the rate of energy transfer slows down as the ice thickens.

Example 6.9

A flask containing some warm water is connected to a vacuum pump. The pressure inside the flask is reduced. Explain
(i) what you will observe happening to the water,
(ii) why the temperature of the water falls. **(6 marks)**

Solution 6.9

(i) Bubbles rise to the surface of the water. The water is boiling.
[Water boils at a lower temperature when the pressure is reduced: see Section 6.4.]

(ii) For the water to change to vapour it must take up its latent heat. It gets this from the water itself.

6.7 Have You Mastered the Basics?

1. Do you understand the meaning of specific heat capacity and specific latent heat?
2. Do you know how impurities and pressure affect the boiling and melting points of liquids and solids?
3. The specific heat capacity of water is 4200 J (kg K). How much heat is needed to raise 2 kg through 5°C?
4. Can you describe how a refrigerator works?
5. The latent heat of vaporisation of water is 2.26×10^6 J/kg. How much heat is needed to change 2 kg of water to steam at 100°C?

6.8 Answers and Hints on Solutions to 'Have You Mastered The Basics?'

1. See Sections 6.2 and 6.3, and Examples 6.1, 6.3, 6.4, 6.5 and 6.6.
2. See Section 6.4.
3. $Q = mcT$ [see Section 6.2] = (2 × 4200 × 5) J = 42 000 J.
4. See Section 6.5 and Example 6.8.
5. $Q = mL$ [see Section 6.3] = (2 × 2.26 × 10^6) J = 4.52 × 10^6 J = 4.52 MJ.

6.9 Questions

(Answers and hints on solutions will be found in Section 6.10.)

Question 6.1

A mass of 0.4 kg of oil in a container is warmed from 20°C to 24°C by 3360 J of energy. The specific heat capacity of the oil, in J/(kg K), is

A 0.4 × 4 × 3360   B $\dfrac{0.4 \times 4}{3360}$ C $\dfrac{0.4 \times 3360}{4}$ D $\dfrac{3360}{0.4 \times 4}$ E $\dfrac{4 \times 3360}{0.4}$ (AEB)

Question 6.2

The water at the foot of a waterfall has a higher temperature than the water at the top even though the surrounding temperatures are equal. What is this temperature difference for a waterfall 100 m in height, given that all the available energy appears as heat in the water? (Assume specific heat capacity of water = 4000 J/kg K, acceleration of free fall = 10 m/s².)

A $\frac{1}{8}$ K   B $\frac{1}{4}$ K   C $\frac{1}{2}$ K   D 2 K   E 4 K   (L)

75

Question 6.3

(a) A 100 W electric immersion heater is placed in a 2.00 kg block of aluminium. If the heater is left on for 5.00 minutes, what will be the temperature rise of the block? (Specific heat capacity of aluminium = 840 J/(kg K).)

(b) (i) Explain a useful application of the fact that water has a high specific heat capacity.
(ii) Why are good fruit growing areas in the world often found near lakes?

Question 6.4

(a) Why is it easier to skate on ice at −2°C than on ice at −20°C?

(b) Explain why it would take much longer to boil an egg at the top of a high mountain than at sea level. **(6 marks)**

Question 6.5

The diagram shows the essential features of a cold store (a room maintained at a low temperature).

(a) Why are the cooling pipes at the top of the cold store? **(2 marks)**

(b) Would it work as well if all the refrigerating machinery were inside the cold store? Explain your answer. **(3 marks)**

(c) If the refrigerating machine were left on continuously, would the cold store go on getting colder and colder? Explain your answer. **(3 marks)**

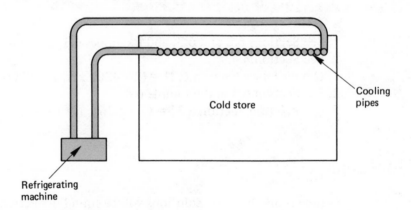

(d) As the liquid refrigerant goes into the cooling pipes, it changes from a liquid to a gas. Discuss this change of state in terms of what is happening to the molecules as they pass into the pipes in the cold store.

Question 6.6

(a) Describe briefly the difference between the behaviour of molecules in the solid, liquid and gaseous state. **(3 marks)**

On the basis of molecular behaviour explain why
(i) energy is needed to change a solid at its melting point into liquid without a change in temperature, **(3 marks)**
(ii) evaporation can take place at temperatures below the boiling point, **(3 marks)**
(iii) the pressure of a constant volume of enclosed gas increases when its temperature is raised. **(4 marks)**

(b)

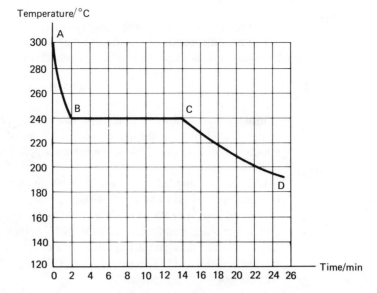

The graph shows the variation of temperature with time for a pure metal cooling from 300°C. In what state is the metal in stage
  (i) AB,
  (ii) BC,
  (iii) CD? **(3 marks)**
If the average rate of heat loss during stage BC is 120 J/min and the mass of metal is 80 g, what is the specific latent heat of fusion of the metal? **(4 marks)**

(L)

Question 6.7

PART I

(a) Define and name the units of
  (i) specific heat capacity,
  (ii) specific latent heat. **(6 marks)**

(b)

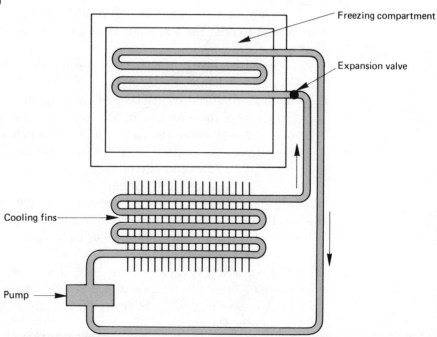

The diagram shows a simple refrigerator and the arrows indicate the direction in which the refrigerant flows. Explain what is happening

(i) in the tubes in the freezing compartment,

(ii) in the tubes connected to the cooling fins.

For each location name the *state* of the refrigerant.   **(6 marks)**

PART II

(c) A manufacturer designs an immersion heater which has a power output of 120 W. The heater is used to raise the temperature of 2 kg of a liquid from 15°C to 35°C in 10 minutes. Assuming that 25% of the energy supplied by the heater is lost from the liquid to the surroundings, calculate

  (i) the energy supplied by the heater in 10 minutes,

 (ii) the specific heat capacity of the liquid.

    The immersion heater is made of insulated wire of resistance 12 Ω/m. If the immersion heater is designed for use with the 240 V main supply, calculate

(iii) the current through the immersion heater,

(iv) the resistance of the immersion heater,

 (v) the length of wire required.   **(19 marks)**

          (AEB)

Question 6.8

PART I

(a) State *two* factors which change the boiling point of water, and in each case state how the boiling point is affected.   **(2 marks)**

(b) Explain the difference between *evaporation* and *boiling*.   **(4 marks)**

          (AEB, part question)

6.10 Answers and Hints on Solutions to Questions

1. <u>Answer **D**</u>

2. Equate the potential energy of 1 kg of water at the top of the waterfall to the heat needed to raise the temperature of 1 kg through t°C. $t = \frac{1}{4}$ K.
<u>Answer **B**</u>

3. Heat supplied by heater = heat taken up by block. $(100 \times 5 \times 60)$ J = 2 kg $\times$ 840 J/(kg K) $\times$ T; hence, $T = 17.8$°C.

4. (a) The heat produced by friction will melt ice at -2°C more easily than ice at -20°C. Also the pressure of the skate on the ice will lower the melting point and may cause the ice to melt. Both mean that there is water under the skate and the skate flows more smoothly over the ice.

    (b) The pressure is less, so the boiling point of water decreases. It takes longer to provide sufficient heat to 'boil' the egg.

5. (a) Cold air is denser than hot air and therefore the cold air sinks.

    (b) No. If the refrigerator machine is inside, the heat withdrawn (by the fluid flowing in the pipes) would be put back again by the refrigerating machine (see Section 6.5).

    (c) As the inside gets colder, more heat flows from the outside to the inside. Eventually a steady temperature is reached when the rate of extraction of heat by the refrigerator is equal to the rate of flow of heat to the inside through the walls.

    (d) The molecules gain enough energy to break away from the attractive forces in the liquid state. Their average kinetic energy has increased and the liquid has become a gas.

6. (a) (i), (ii) and (iii) — see Section 5.4.
   (b) (i) liquid; (ii) liquid and solid present together — i.e. the liquid is solidifying; (iii) solid.

   Use $Q = mL$ [Section 6.3.] $\{120 \times (14 - 2)\}$ J = 0.080 kg $\times$ L; hence, $L = 1.8 \times 10^4$ J/kg.

7. (a) See Sections 6.2 and 6.3.
   (b) (i) The refrigerant passes onto the low-pressure side of the system, becomes a gas and cools. (ii) The gas under pressure liquefies and gives out its latent heat.
   (c) (i) 72 000 J. (ii) 54 000 J = 2 kg $\times$ c $\times$ 20 K; hence, c = 1350 J/(kg K). (iii) Use the equation in Section 10.2. Current = 0.5 A. (iv) Use equation in Section 10.1. Resistance = 480 Ω. (v) Length of wire = 40 m.

8. (a) See Section 6.4.
   (b) See Section 5.4.

7 Transfer of Heat, Convection, Conduction, Radiation, the Thermos Flask and the Greenhouse Effect

7.1 Heat Transfer

Heat energy may be transferred from one place to another by conduction, convection and radiation.

7.2 Conduction

Conduction is the flow of heat energy through a body which is not at uniform temperature, from places of higher temperature to places of lower temperature, without the body as a whole moving.

If a metal rod is placed on a tripod and one end heated with a bunsen burner, then heat flows down the rod by the process of conduction and the other end begins to warm up. The energy is passed down the rod by the free electrons in the metal and also by the vibrating atoms passing on their energy to adjacent atoms.

Bad conductors are used to insulate roofs, water pipes and storage tanks, and for the handles of saucepans and teapots. Air is a poor conductor of heat, and cellular blankets, string vests, fibre glass and fur coats depend for their insulating properties on the air which is trapped in them. Good conductors are used as bases for saucepans and for cooling fins in air cooled engines.

The rate of conduction of heat through a material depends on (a) its nature, (b) its thickness, (c) its area and (d) the temperature difference across its thickness.

7.3 Convection

Convection is the transfer of heat energy by the circulation of a fluid (a liquid or a gas) due to temperature difference within it.

The essential difference between convection and conduction is that in convection the less dense hot body rises and takes its heat with it, and in conduction heat flows through the body, which does not move. Hot water systems, coastal breezes from land to sea at night and the hot air rising above radiators are examples of convection.

7.4 Radiation

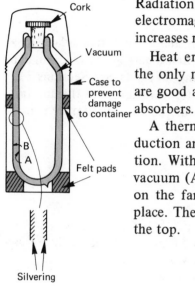

A vacuum flask

Radiation is the transfer of heat energy from one place to another by means of electromagnetic waves. The amount of heat energy radiated per second by a body increases rapidly as its temperature rises.

Heat energy from the Sun reaches us by electromagnetic waves. Radiation is the only means of heat transfer which can take place in a vacuum. Black surfaces are good absorbers and good radiators. Shiny surfaces are poor radiators and poor absorbers.

A thermos (vacuum) flask is designed to reduce heat flow by convection, conduction and radiation. The vacuum prevents heat flow by conduction and convection. With a hot liquid in the flask the shiny surface on the liquid side of the vacuum (A) is a poor radiator and reduces heat loss by radiation. The shiny surface on the far side (B) reflects back the small amount of radiation that does take place. The main loss of heat is by conduction up the sides and through the cork at the top.

7.5 House Insulation

Fibre glass or other insulators in the walls and roofs of houses considerably reduce heat loss by conduction and also by convection. Not only does double glazing produce another layer of poorly conducting glass, but also the air trapped between the glass surfaces is a bad conductor of heat. Plastic strips, coated with shiny aluminium, fitted behind radiators reflect the infra-red radiation back into the room.

7.6 The Greenhouse Effect

The shorter wavelength infra-red radiation and visible light from the Sun pass through the glass of a greenhouse and are absorbed by the soil and the plants, raising the temperature. The infra-red radiation emitted by the soil and plants is of much longer wavelength than the infra-red radiation emitted by the Sun (the higher the temperature the shorter is the wavelength of the radiation emitted). The longer wavelength infra-red radiation does not pass through the glass and so energy is not lost from the greenhouse; it is trapped in the greenhouse and the greenhouse temperature rises.

7.7 Worked Examples

Example 7.1

Which of the following statements about heat transfer is correct?
A Conduction takes place both in liquids and in a vacuum
B Convection takes place in both liquids and solids
C The amount of heat energy radiated every second by a body increases slowly as its temperature rises
D Radiation is the *only* way heat transfer can take place in a vacuum

Solution 7.1

[Convection and conduction both need a medium and cannot take place in a vacuum. Convection requires that the warmer, less dense part of the medium can rise and take its heat with it, so it can take place only in fluids. See Sections 7.2, 7.3 and 7.4. Very little radiation takes place at low temperatures, but the amount of heat radiated from a body does increase *very rapidly* as the temperature rises.]

Answer **D**

Example 7.2

The windows of many modern buildings are 'double-glazed' (i.e. have two thicknesses of glass with a small air space between) to reduce heat losses to the outside. This is mainly because
A evaporation of moisture from the outside of the windows is reduced
B convection currents cannot pass through the extra layer of glass
C radiated heat is not transmitted through the air space
D air is a very bad conductor of heat
E glass is a very bad conductor of heat

(AEB)

Solution 7.2

[The air trapped between the sheets of glass is a bad conductor of heat and this reduces the heat loss from the house. It is also true that glass is a bad conductor and this fact also reduces the heat loss, but the *main* reason is the poor conductivity of air. A thick sheet of glass would be cheaper but not so effective.]

Answer **D**

Example 7.3

In cold weather the metal blade of a knife feels colder than the wooden handle because the
A metal is at a lower temperature than the wood
B metal is a better conductor of heat than the wood
C metal has a smaller specific heat capacity than the wood
D metal has a brighter surface than the wood
E molecules in the metal are vibrating more vigorously than those in the wood   (AEB)

Solution 7.3

[Because the metal is a good conductor of heat, when it comes into contact with skin the heat from the skin can easily pass through the metal blade. Wood is not such a good conductor and the heat cannot easily pass from the hand to the wood.]

Answer **B**

Example 7.4

A person sitting on a beach on a calm hot summer's day is aware of a cool breeze blowing from the sea. Explain why there is a breeze. **(6 marks)**

(L)

Solution 7.4

The land has a lower specific heat capacity than the sea and in daytime the land is hotter than the sea. The air above the land is warmer and less dense than the air over the sea, and it rises, the cold air from the sea coming in to take its place.

Example 7.5

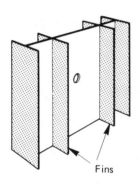

Fins

(a) Heat energy may be transferred by conduction, convection and radiation. Describe three experiments, one for each process, to illustrate the methods of heat transfer. **(12 marks)**

(b) Some electrical devices, such as power transistors, can become so hot that they do not function properly. In order to prevent this they are fastened in good thermal contact with a 'heat sink', such as a piece of aluminium sheet with aluminium fins as shown in the diagram.

  (i) What is meant by 'good thermal contact'?

  (ii) Explain how the heat is carried away from the electrical device to the air outside it.

  (iii) Why does the heat sink have fins?

  (iv) Discuss whether the heat sink would operate better if it were placed with its fins horizontal, rather than vertical as shown in the diagram. **(8 marks)**

Solution 7.5

(a) A metal rod is placed on a tripod. The rod should be long enough to ensure both ends are well clear of the tripod. A bunsen burner is placed under one end. Very soon the other end becomes warm; its temperature rises. This can be detected simply by feeling the end of the rod. Heat has been passed down the rod by conduction.

Convection may be demonstrated using the apparatus shown in the diagram. A box has two chimneys as shown. A lighted candle is placed under one of the chimneys and the glass front closed. Very soon smoke from the smouldering rope will be seen passing through the box and out of the chimney above the candle. Convection currents of air are passing through the box.

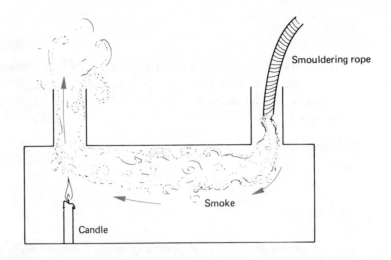

Sit in front of an electric fire which has a shiny reflector behind it. You will feel the radiation reflected on to your body. The heat has not arrived by conduction because air is a bad conductor. Hot air convection currents will rise above the element and circulate around the room but this is not how most of the heat reaches your body. Most of it has arrived as a result of radiation from the element falling on your body.

(b) (i) Contact so that heat can flow from one body to the other. Some air between the two bodies would prevent good thermal contact.

(ii) Heat is conducted from the device along the aluminium sheet and through the fins. The air in contact with the aluminium and fins becomes warm and the less dense warm air rises, carrying its heat with it. Denser colder air replaces the warm air and the process continues. At low temperatures very little heat is lost by radiation.

(iii) The fins increase the area of surface in contact with the air, thus increasing the heat loss from the metal surfaces.

(iv) It operates better if placed vertically because the convection currents can flow more freely.

Example 7.6

(a)

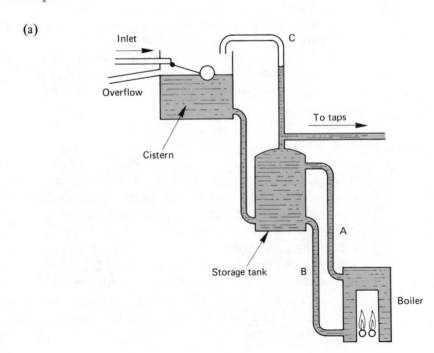

84

The diagram shows part of a household hot water system.

(i) Why is pipe A connected between the top of the boiler and the top of the storage tank? **(3 marks)**

(ii) Why is pipe B connected between the bottom of the boiler and the bottom of the storage tank? **(3 marks)**

(iii) What is the function of pipe C? **(2 marks)**

(iv) Suggest, with reasons, what might be added to the hot water system above to make it more efficient. **(3 marks)**

(b) The temperature of the water inside an aquarium can be controlled by a thermostat which switches an electric heater on and off. Draw a diagram showing how this may be done using a bimetallic strip. (Your diagram must clearly show the construction of the bimetallic strip.)

How may different constant temperatures be achieved using your arrangement?

(5 marks)

(L, part question)

Solution 7.6

(a) (i) and (ii)

The hot, less dense water in the boiler rises and goes via the pipe A to the top of the storage tank. The denser, colder water from the bottom of the storage tank passes into the boiler. The pipe A is connected to the top of both tanks because hot water is less dense than cold water, and pipe B is connected to the bottom of both tanks because cold water is more dense than hot water.

(iii) C is an expansion pipe. It is a safety precaution to allow steam to escape should the water boil. It also allows any dissolved air which is released from the heated water to escape, thus helping to prevent air locks.

(iv) Lagging should be added round all the hot pipes and round the hot water tank. This will reduce the heat lost to the atmosphere and make the system more efficient.

(b) If the screw is screwed down the thermostat will switch off at a higher temperature.

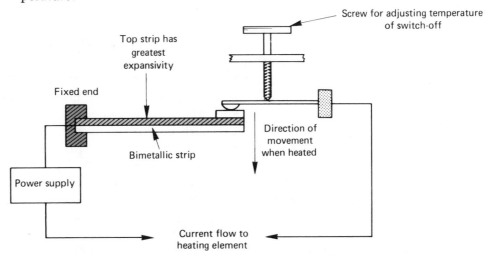

[A similar thermostat has many uses — i.e. in domestic boilers and ovens.]

Example 7.7

PART I

(a) How would you demonstrate that water is a poor conductor of heat? **(4 marks)**

(b) Explain how energy is transferred from the heating element throughout the water in an electric kettle. **(4 marks)**

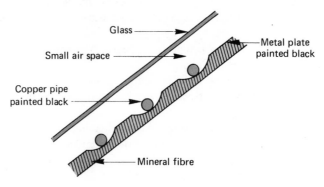

Glass

Small air space

Metal plate
painted black

Copper pipe
painted black

Mineral fibre

Construction of the panel

PART II

Start a new page

(c) The diagram shows the essential features of a solar heating panel. A small electric pump circulates a liquid through the pipes. State briefly why
  (i) the pipes and back plate are blackened
  (ii) there is a mineral fibre backing to the panel
  (iii) the glass sheet increases the energy collected by the panel by a large factor.
  (5 marks)

(d) Each day the solar panel raises the temperature of 200 kg of water from 5°C to 35°C. Calculate the energy incident each day on the panel if only 20% of this energy is absorbed by the panel. **(5 marks)**

(e) If no solar panel were available an electric immersion heater could be used. What would be the daily cost if the charge for electricity were 5p per kW h? (Neglect heat lost to the surroundings.) **(3 marks)**

Data for this question

Specific heat capacity of water = 4200 J/(kg K). (AEB)

Solution 7.7

(a) Put some ice wrapped in a metal gauze (so that it sinks) in a test tube which is nearly full of water. Carefully heat the water at the top of the tube until it boils. The water and ice at the bottom of the tube remain cold because the water (and of course the glass) are poor conductors of heat [a diagram of the experiment is in Question 7.7 at the end of this chapter].

(b) The water in contact with the heating element gets hotter. Hot water is less dense than cold water, so the hot water rises. The denser surrounding cold water flows in to take its place. These circulating convection currents gradually transfer the heat throughout the water in the kettle.

(c) (i) Black surfaces are better absorbers of heat than shiny or light-coloured surfaces. The solar panel works better if as much heat as possible is absorbed from the Sun's rays.

  (ii) The mineral fibre is a poor conductor and this reduces the heat escaping from the copper pipes.

  (iii) The electromagnetic waves emitted by the copper pipes are of much longer wavelength than the rays arriving from the Sun. They are not transmitted by the glass.

(d) Energy gained by water = mcT [see Section 6.2] = $(200 \text{ kg}) \left(4200 \dfrac{J}{\text{kg K}} \right) (30 \text{ K}$

$$= 2.5 \times 10^7 \text{ J}$$
$$= 25 \text{ MJ}$$

86

This is 20% or $\frac{1}{5}$ of the incident energy.

⇒ energy incident each day = (5 × 25) MJ

= 125 MJ

(e) $1 \text{ kWh} = \left(1000 \, \frac{J}{s}\right) (3600 \text{ s}) = 3.6 \text{ MJ}$

3.6 MJ costs 5 p.

∴ 25 MJ costs $\frac{5 \times 25}{3.6}$ p = 35 p

Cost is 35 p.

[The temperature change, 30 K, is only known to two significant figures, so the answer may only be given to two significant figures.]

7.8 Have You Mastered the Basics?

1. Can you distinguish between convection, conduction and radiation?
2. Can you describe experiments to illustrate each of them?
3. Can you explain how a vacuum flask works?
4. Do you know what surfaces are the best radiators and absorbers of heat?
5. Can you explain how a household hot water system works?
6. Can you explain the principles of house insulation?

7.9 Answers to 'Have You Mastered the Basics?'

1. See Sections 7.2, 7.3 and 7.4.
2. See Sections 7.2 and 7.3 and Example 7.5.
3. See Section 7.4.
4. See Section 7.4.
5. See Example 7.6.
6. See Section 7.5.

7.10 Questions

(Answers and hints on solutions will be found in Section 7.11.)

Questions 7.1–7.3

Some ways in which heat may be transferred are

A Conduction only

B Convection only

C Radiation only

D Convection and conduction

E Conduction and radiation

In each of the following three questions choose from **A** to **E** above and select the one which best describes the process or processes by which heat is transmitted.

1. From the Sun to the Earth

2. From the hot metal element of an electric kettle, which is in contact with the water, to the rest of the water in the kettle

3. From the inside glass surface of a vacuum flask to the outside of the outermost glass surface

Question 7.4

Explain each of the following:
(i) A candle placed at the bottom of a gas jar soon goes out, but if a partition is carefully lowered down the middle of the jar so that the top half of the jar has a partition in it, the candle will stay alight.
(ii) Many good insulators of heat are porous materials.
(iii) A greenhouse without artificial heat is warmer inside than the air temperature outside.

Question 7.5

(a) (i) 'The *double walls* of a *vacuum* flask are *silvered* to reduce heat transfer between the contents of the flask and its surroundings.' Explain the purpose of each of the three features in italics in this statement. **(4 marks)**
    (ii) List the possible reasons why the contents of a well-sealed vacuum flask eventually reach the same temperature as that of its surroundings. **(3 marks)**
(b) In coastal regions during periods of hot sunny weather a steady light wind, reversing its direction every twelve hours, is often noticed. Account for the existence of the *day-time* breeze and with the help of a diagram show the direction of the air currents over the land. **(7 marks)**
(c) Many householders in Great Britain have built 'sun porches' (rooms with very large windows) on the south side of their houses. Explain, with the aid of a diagram, why the day-time temperature of these rooms is always higher than that of the air outside.
    (6 marks)
    (L)

Question 7.6

(a) Heat energy may be transferred from one body to another by convection, conduction and radiation. Describe how the heat energy is transferred in each case. **(10 marks)**
(b) Describe three experiments, one for each process, to illustrate the methods of transfer.
    (9 marks)
(c) Explain the part played by each of the three processes in heating a room using a hot water radiator. **(6 marks)**

Question 7.7

Water
Heat
Ice weighted with copper gauze

(a) Describe an experiment which you would carry out to show how the nature of a surface affects the heat radiated from that surface in a given time. **(5 marks)**
    State any precautions which you would take and state your findings for two named surfaces. **(3 marks)**
    How would you then show that the surface which was the better radiator was also the better absorber of radiation? **(4 marks)**
(b) As the surface of a pond freezes it is found that each equal increase in the thickness of the ice takes longer to form, even when the air above the ice remains at the same low temperature. Explain why this is so. **(4 marks)**
(c) In the experiment shown the ice remains intact for several minutes as heating progresses. Explain how this can be so. **(4 marks)**
    (L)

7.11 Answers and Hints on Solutions to Questions

1. **C**

2. **D** [mainly convection but water does conduct (though poorly)].

3. Conduction through the glass and radiation across the vacuum.
   <u>Answer E</u>

4.   (i) Warm air rises up one side of the partition and cold air comes down the other side, supplying oxygen to the flame.
    (ii) The trapped air is a poor conductor.
   (iii) See Section 7.6.

5. (a) See Section 7.4. There must also be a very small loss due to radiation across the vacuum.
   (b) See Example 7.4.
   (c) This is the greenhouse effect; see Section 7.6.

6. (a) In convection the warm fluid rises. The average kinetic energy of the molecules in the warm fluid is higher than that of those in the colder parts of the fluid. The energy is thus transported by the moving fluid. In conduction the molecules at the hotter end are vibrating with greater energy than the molecules further down the body. They jostle the molecules near them and pass on their energy through the body. In metals, the free electrons at the hot end have a greater kinetic energy than the electrons at the cold end. They 'wander about' in the metal and transfer their energy to the colder parts of the metal. In radiation the energy is carried by the electromagnetic wave.
   (b) See Example 7.5.
   (c) Conduction through the metal of the radiator, convection currents in the air and a small amount of radiation.

7. (a) Heat a copper plate which is shiny on one side and blackened on the other. Clamp the plate vertically and place your cheek near to each side in turn. Alternatively detect the heat with a thermopile (a detector of radiant heat; the deflection of the galvanometer attached to it is a measure of the heat entering the thermopile). You must ensure that the heat reaching the detector comes from the emitting surface. Black surfaces are much better radiators than shiny surfaces. Place two surfaces, one blackened, with thermometers in good thermal contact with their surfaces, at equal distances from a cylindrical heating element. The thermometer on the black surface rises faster than the one on the shiny surface.
   (b) Ice is a bad conductor of heat.
   (c) The water and glass are bad conductors of heat.

8 Reflection, Refraction, the Electromagnetic Spectrum, Lenses and Curved Mirrors

8.1 Law of Reflection

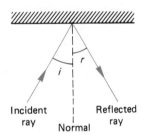

Figure 8.1 The angle of incidence i is equal to the angle of reflection r.

The angle of incidence is equal to the angle of reflection (Fig. 8.1).

8.2 Plane Mirrors

The image formed by a plane mirror lies on the normal from the object to the mirror and is as far behind the mirror as the object is in front (Fig. 8.2). The image is virtual, laterally inverted and the same size as the object.

A simple periscope may be constructed using two reflecting prisms or two plane mirrors (see Example 8.10).

Plane mirrors are often placed behind a scale over which a pointer passes. By aligning the eye with the pointer and its image the error due to parallax in reading the scale is avoided.

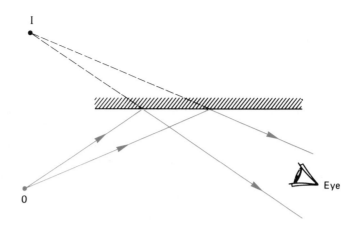

Figure 8.2 The image lies on the perpendicular bisector of the mirror and is as far behind the mirror as the object is in front.

8.3 Refraction

When light (or other wave motion) crosses a boundary between two different media, it is refracted. Refraction results from the change in speed of the wave as it crosses the boundary. The greater the change in the speed the greater is the refraction of the light.

If an object under water or under a glass block is viewed vertically from above then the apparent depth is less than the real depth (see Example 8.6).

When a ray of white light falls on a prism, the different colours composing the white light are each refracted a different amount and a spectrum is formed. The wavelength of red light is greater than the wavelength of blue light. The greater the amplitude of the waves the greater the brightness of the light.

8.4 Critical Angle, Prisms, Optical Fibres

The critical angle for any medium is the angle of incidence of light on the boundary such that the angle of refraction is 90° (Fig. 8.3). If the angle of incidence is greater than the critical angle, total internal reflection occurs. Total internal reflection can only occur when light is passing from an optically denser to an optically less dense medium (i.e. from glass to air).

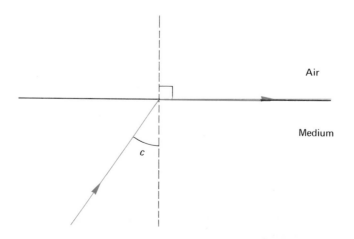

Figure 8.3 A ray of light incident at the critical angle.

Isosceles totally internally reflecting prisms are often used instead of plane mirrors because (i) they do not form multiple images (see Example 8.4), and (ii) there is no silvering to wear off.

Optical fibres make use of total internal reflection. The outside 'cladding' is less dense than the core, and electromagnetic waves travelling along the fibre are continually being totally internally reflected (in a similar way to the light in a plastic tube: see Example 8.7). Optical fibres are now being used instead of wires for telephone cables. The fibres can carry more messages and it is less easy to interfere with the transmission, because it travels down the centre of the tube. Security is therefore greater. Signals in optical fibres (usually infra-red laser light) stay strong over larger distances and so 'boosters' can be a long way apart.

8.5 The Electromagnetic Spectrum

Electromagnetic waves are characterised by oscillating electric and magnetic fields. In a vacuum they all travel at the speed of light. The spectrum of these waves includes γ-rays, X-rays, ultra-violet rays, visible light, infra-red rays, radar waves and radio waves. The list is in order of increasing wavelength (decreasing frequency), γ-rays having the shortest wavelength (largest frequency).

Electromagnetic waves are used in (a) communication systems (links between satellites and ground stations usually use microwaves, and telephone systems use infra-red laser light passing through optical fibres); (b) microwave ovens (the microwaves penetrate the food and are absorbed by water molecules in the food, thus heating the food from within. Microwaves pass through air, glass and plastic without causing any heating and are reflected by metals); and (c) security systems (the breaking of an invisible infra-red beam can be used to trigger an alarm system).

8.6 Lenses

The principal terms used to describe the action of a lens are shown in Fig. 8.4. Parallel rays converge to a point in the *focal plane*. The distance from the focal plane to the lens is the *focal length* of the lens. If a lens is used to focus a distant object on a screen, then the distance from the lens to the screen is the focal length.

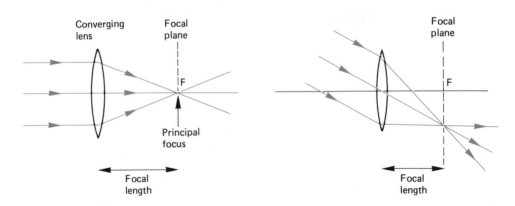

Figure 8.4 Parallel rays arriving from a distant object and passing through a converging lens.

All objects which are a long way from the lens form inverted real images close to the focal plane, and this fact is made use of in the *camera*.

If the object is close to the principal focus but outside it, the image formed is inverted, real and magnified. A lens is used in this way in a *slide projector*.

If the object is at a distance from the lens which is less than the focal length of the lens, then a virtual, magnified, erect image is formed. When used in this way, the lens acts as a magnifying glass.

These facts are summarised in Table 8.1.

$$\text{Magnification} = \frac{\text{height of image}}{\text{height of object}} = \frac{\text{image distance}}{\text{object distance}}$$

Table 8.1   Images formed by a converging lens

Object position	Image position	Use
At infinity, or a very large distance	At F, or very close to F	
Outside F but close to it	Quite large distance from lens	
Inside F	Same side of lens as object but further from lens	

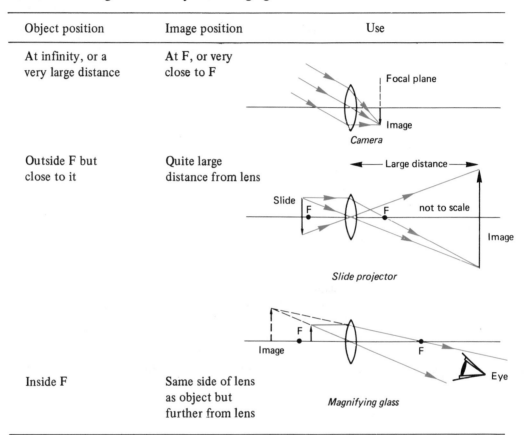

8.7 **The Eye, the Camera and the Slide Projector**

The eye and the camera both have a converging lens which forms a real, diminished, inverted image of the object on a light sensitive area (the *retina* in the eye, the film in a camera). *A slide projector* has a converging *projection lens* which forms a real, inverted, magnified image on a screen.

8.8 **Curved Mirrors**

Parallel beams of light incident on a concave mirror are focused at the principal focus (see Fig. 8.5). This property of concave mirrors is made use of in microwave dishes and radio telescopes: the waves from distant sources are brought to a focus. In headlamp reflectors the source is placed at the principal focus and a parallel beam is produced. Electric fires also make use of curved reflectors.

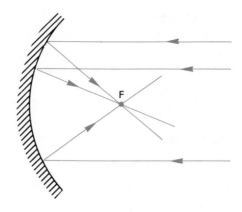

Figure 8.5 The curved reflector brings parallel rays to a focus at F. On the other hand, if a source is placed at F, a parallel beam results.

8.9 Worked Examples

Example 8.1

Which one of the following is not a property of the image of an object placed 12 cm in front of a plane mirror?

A It is behind the mirror
B It is 12 cm from the mirror
C It is laterally inverted
D It is real
E A line joining the top of the object to the top of the image is perpendicular to the plane of the mirror

Solution 8.1

[No light travels from the object to points behind the mirror. The rays of light reflected at the mirror *appear* to come from a point behind the mirror. The image is virtual.]

Answer **D**

Example 8.2

(i) A girl stands at a distance of 2 m in front of a plane mirror, and a boy stands at a distance of 3 m in front of the same mirror. How far from the boy is the girl's image in the mirror? **(3 marks)**

(ii) The girl stands still and the mirror is moved away from her at 3 m/s. At one instant the girl and her image are 6 m apart. How far apart will they be 2 s later? **(5 marks)**

Solution 8.2

(i) The girl's image is 2 m behind the mirror. The boy is 3 m in front of the mirror. Therefore the girl's image is 5 m from the boy.

(ii) When they are 6 m apart the girl is 3 m in front of the mirror and her image is 3 m behind the mirror. 2 s later she is 9 m in front of the mirror and her image is 9 m behind the mirror. They are 18 m apart.

Example 8.3

(a) A pin is placed in front of a plane mirror. Describe an experiment you would do to locate the position of the pin's image.   **(9 marks)**

(b) In the diagram the straight line HEF represents a girl standing in front of a plane mirror. H is the top of her head, E her eyes and F her feet. The girl is 140 cm high and her eyes are 10 cm below the top of her head. Draw a ray diagram (which need not be to scale) showing
   (i) a ray of light which travels from the top of her head to her eyes and
   (ii) a ray of light which travels from her feet to her eyes.
(c) What is the minimum length of the mirror that would be required in order to enable her to see a full length image of herself in the mirror?   **(11 marks)**

Solution 8.3

(a) The plane mirror is placed on a sheet of paper with its surface vertical and the object pin O placed in front of the mirror. With the eye at A two pins A_1 and

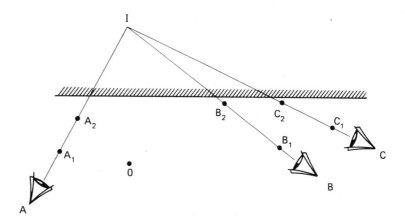

A_2 are placed so that they are in line with the image I. The procedure is repeated with the eye at B and C. The mirror is removed and the lines $A_1 A_2$, $B_1 B_2$ and $C_1 C_2$ are drawn. They intersect at the position of the image I.

95

(b)

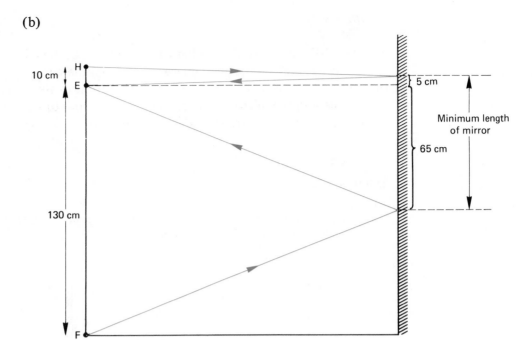

(c) As shown in the diagram the minimum length of the mirror is 70 cm.

Example 8.4

An object is placed in front of a thick sheet of glass.

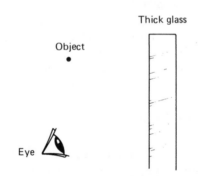

(a) An image, I_1, is formed by reflection from the front surface of the glass.
   (i) Mark and label the exact position of I_1.
   (ii) Draw a ray diagram to show how this image is seen by the eye.
(b) A second image, I_2, will also be seen by the same observer.
   (i) What causes the formation of this second image?
   (ii) Mark on the diagram the position of this second image, I_2.   (4 marks)
                                                                                    (AEB)

Solution 8.4

(a) [On the scale of the diagram O is 2.2 cm in front of near surface of glass, so I_1 is 2.2 cm behind near surface of glass. Draw lines from either side of the pupil directed towards I_1. When they hit the near surface they go to the object. Don't forget to dot the virtual rays and to put arrows on light rays from the object going into the eye.]

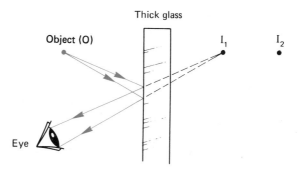

Thick glass

Object (O) I_1 I_2

Eye

(b) (i) Reflection at the rear surface of the glass of light refracted at the first interface.

(ii) [The object is 3 cm in front of this rear surface and the image I_2 is nearly 3 cm behind it. It would be exactly 3 cm if there were no refraction at the front surface of the glass.]

Example 8.5

Which one of the diagrams correctly shows the path of the ray through the glass block?

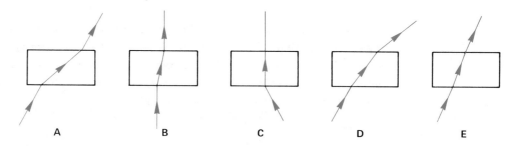

A   B   C   D   E

Solution 8.5

[When light passes from air to glass it is bent towards the normal. It is bent away from the normal when it passes from glass to air. It emerges parallel to the incident ray.]

<u>Answer E</u>

Example 8.6

(a) When you look vertically down into a pond, the pond appears to be shallower than it really is. Draw a ray diagram to illustrate this phenomenon. Mark clearly on your diagram the position of the bottom of the pond and where the bottom appears to be. **(5 marks)**

(b) The diagrams show a ray of light incident on a glass prism. In each case complete the diagrams, showing the subsequent path of the ray. **(5 marks)**

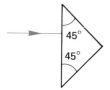

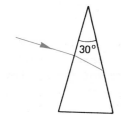

Solution 8.6

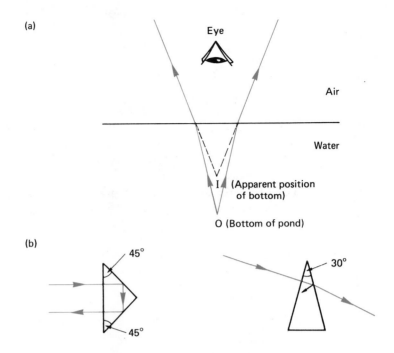

(a)

Eye

Air

Water

I (Apparent position of bottom)

O (Bottom of pond)

(b)

45°

45°

30°

[For (b) in the first diagram the ray is incident on the glass–air interface at an angle greater than the critical angle and is totally internally reflected. In the second diagram the angle of incidence on the glass–air interface is much less than the critical angle (the critical angle is about $42°$ for glass) and the ray passes into the air, being bent away from the normal.]

Example 8.7

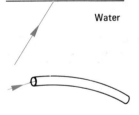

Air

Water

(a) The diagram shows a ray of light in water incident on a water/air boundary. Draw sketches to show what would happen to the ray of light when the angle of incidence is
   (i) about $15°$
   (ii) about $60°$. **(5 marks)**

(b) The diagram shows a light beam incident on a curved transparent plastic tube. Explain why the light will stay in the tube and come out at the other end. **(4 marks)**

Solution 8.7

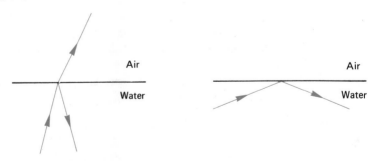

Air

Water

Air

Water

(a) [When the angle of incidence is $15°$ the light passes into the air, being refracted away from the normal. Some light will be reflected. At an angle of incidence of $60°$, total internal reflection occurs and no light passes into the air. Total internal reflection occurs whenever the angle of incidence in the glass exceeds $42°$.]

(b)

The diagram shows the path of the ray. Each time it is incident on the plastic/air surface the angle of incidence is greater than the critical angle and it is totally internally reflected.

Example 8.8

The diagrams show a ray of light incident on a glass prism.

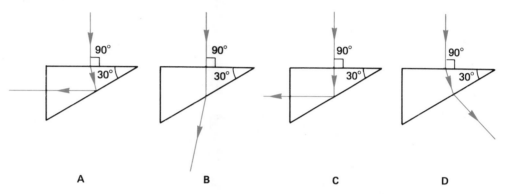

Which diagram shows the subsequent path of the ray through the glass prism?

Solution 8.8

[The ray is incident normally on the block (i.e. the angle of incidence is 0°). It therefore passes into the block undeviated. When it reaches the glass–air interface, it is bent away from the normal, as shown in **B**. It is not totally internally reflected, as the angle of incidence at the glass–air interface is less than the critical angle (the critical angle for glass is about 42°).]
Answer **B**

Example 8.9

(i) Draw a diagram showing how a ray of white light entering a 60° glass prism is refracted and dispersed. **(5 marks)**
(ii) Beyond the red end of the spectrum there is some invisible radiation. How would you detect this radiation? **(6 marks)**
(iii) State, putting them in order of increasing wavelength, five regions of the electromagnetic spectrum. **(6 marks)**

Solution 8.9

(i)

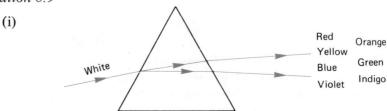

(ii) Infra-red radiation may be detected by using a phototransistor connected in series with a battery and a milliammeter. The reading on the milliammeter is a measure of the infra-red radiation falling on the phototransistor. The phototransistor is moved through the spectrum and into the region beyond the red end. In this position the reading on the milliammeter increases.

(iii) Gamma rays, X-rays, ultra-violet light, infra-red light and radio waves (see Section 8.5).

Example 8.10

Which one of the five diagrams below best represents the path of a ray of light through a periscope?

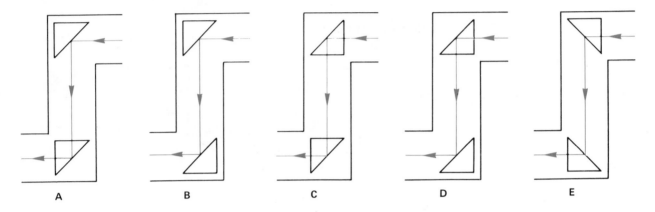

A   B   C   D   E

Solution 8.10

[Prisms are used to reflect light by making use of total internal reflection (see Section 8.4).]

<u>Answer C</u>

Example 8.11

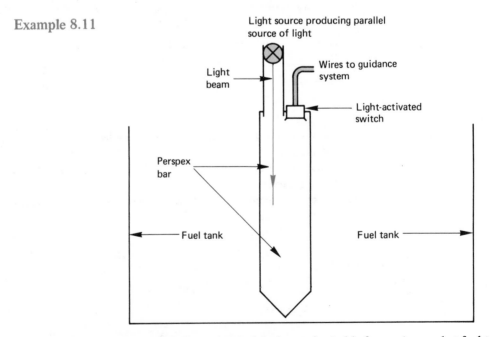

The diagram shows a design for an 'optical fuel gauge' suitable for use in a rocket fuel tank. When the tank runs low, the light-activated switch is triggered.

Air is less 'optically dense' than Perspex. Perspex is less 'optically dense' than rocket fuel.
(i) Complete the diagram showing the paths of the light rays when the tank is
    (a) FULL;   (b) EMPTY.   Label the rays you draw, (a) and (b).   (5 marks)
(ii) Why is the term 'CRITICAL ANGLE' of importance in this application?   (3 marks)
(iii) What is the term given to the process by which light changes direction in order to reach
    the light-activated switch.   (1 mark)

Solution 8.11

(i)

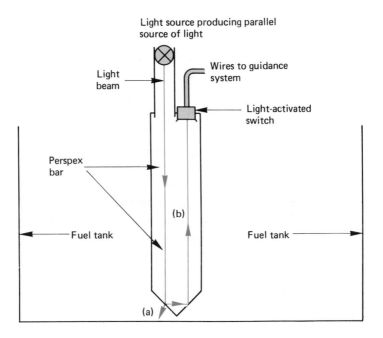

(ii) When light passes from a denser to a less dense medium and the angle of
    incidence in the denser medium is greater than the critical angle, the ray is
    totally internally reflected as shown above, and the light-activated switch is
    triggered.
    [When there is fuel round the Perspex the light is passing into a denser
    medium and is bent towards the normal as shown in (a) above.]
(iii) Total internal reflection.

Example 8.12

(a) The diagram shows an object OO′ placed in front of a converging lens. The image is
    formed on the line XY.

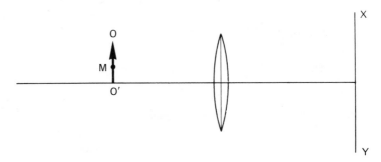

(i) Draw *three* rays leaving O (the arrow head at the top of the object) and passing through the lens. Show clearly the position of the image. **(6 marks)**

(ii) Show the path of two rays leaving M (the mid-point of OO′), which pass through the lens and travel to the image. **(3 marks)**

(b) Suppose the lens were dropped and broken into two approximately equal pieces. What effect, if any, would this have on the brightness and size of the image formed in (a)(i). **(3 marks)**

Solution 8.12

(a) (i) and (ii) (see diagram).

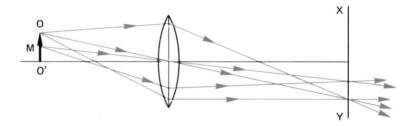

[The ray through the centre of the lens passes through undeviated. Where this meets the line XY is the position of the image. Once this position is fixed any other ray leaving O goes through this same point on the image. All rays leaving M go through the mid-point of the image.]

(b) The rays of light passing through the remaining half will travel the same path as when the whole of the lens was present. The image will therefore be the same size. Since only half the light will reach the image, the image will be less bright.

Example 8.13

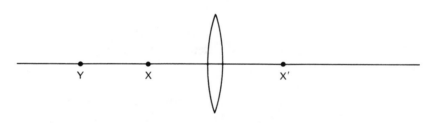

The diagram shows a point object X on the principal axis of a converging lens. The image is formed at X′.

(i) Draw three rays from Y, another point on the principal of the lens, and show a possible path for each of these rays after they pass through the lens. Mark the position of the image Y′. **(4 marks)**

(ii) Draw another diagram showing the paths of rays coming from a distant object and passing through the lens. **(3 marks)**

Solution 8.13

(i)

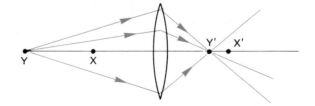

(ii)

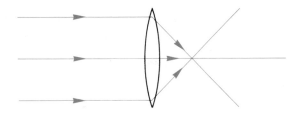

[As the object distance increases, the image distance decreases. When the object is a long way from the lens, the rays arrive at the lens nearly parallel. They pass through the principal focus, which is closer to the lens than Y′.]

Example 8.14

A camera has a lens of focal length 50 mm.
(a) When one takes a photograph of a distant object, where should the film be placed?
                                                                                    (2 marks)
(b) What adjustments can be made to adapt the camera to take photographs in bright sunlight?   (4 marks)

Solution 8.14

(a) Rays from a distant object are brought to a focus in the focal plane of the lens and the film must therefore be placed at the focal plane of the lens – i.e. 50 mm from the lens.
(b) The amount of light entering the camera can be reduced. This may be done by (i) decreasing the aperture or (ii) decreasing the exposure time. Another way is to use a film of lower sensitivity.

Example 8.15

(a) State *one* similarity and *one* difference in the images formed by a slide projector and a camera.   (2 marks)
(b) The diagram shows a camera being used to photograph a distant object. Complete the diagram by continuing the rays to show how the image is formed.   (2 marks)
(c) The same camera is now used to photograph an object which is about 3 m from the lens. What adjustment must be made to the lens in order that a sharp image is formed on the film?   (2 marks)

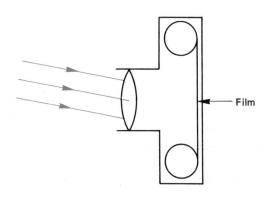

Film

Solution 8.15

(a) Both images are inverted. [Remember that the slide is put in the projector up-side down.] The slide projector image is magnified. The camera image is diminished.

(b)

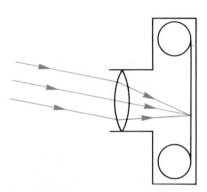

(c) The lens must be moved further from the film. [If the lens were left in the same position, the image would be formed behind the film, so the rays reaching the film would not be focused and a blurred image would be formed on the film.]

8.10 Have You Mastered the Basics?

1. Do you understand the law of reflection? Can you describe an experiment to illustrate it?
2. Can you describe experiments and draw diagrams to illustrate refraction, total internal reflection, critical angle, real and apparent depth, and dispersion?
3. Do you understand the periscope and optical fibres?
4. Can you describe how to produce a spectrum?
5. Can you list the main parts of the electromagnetic spectrum in order of increasing wavelength?
6. Can you explain the terms *principal focus* and *focal length*?
7. Can you explain the action of a converging lens in a camera, in a slide projector and as a magnifying glass?
8. Do you know how to measure the focal length of a lens?

8.11 Answers to 'Have You Mastered the Basics?'

1. See Section 8.1, Fig. 8.1. An experiment could be done with a beam of light incident on a mirror as shown in the diagram.
2. Use beams of light and diagrams as shown in Example 8.5, Section 8.4 (Fig. 8.3), and Examples 8.6 to 8.9.
3. See Sections 8.2 and 8.4 and Example 8.7.
4. See Example 8.9.
5. See Section 8.5.
6. See Section 8.6.
7. See Sections 8.6 and 8.7, Examples 8.14 and 8.15 and Question 8.7.
8. See Section 8.6.

8.12 Questions

(Answers and hints on solutions will be found in Section 8.13.)

Question 8.1

A man whose eyes are 1.50 m from the ground looks at his reflection in a vertical plane mirror 2.00 m away. The top and bottom of the mirror are 2.00 m and 1.00 m from the ground respectively. What distance, in m, below his eyes can the man see of himself?

A 0.25   B 0.50   C 0.75   D 1.00   E 1.50   (AEB)

Question 8.2

(i) Draw a diagram showing a ray of light being reflected from a plane mirror. Mark on your diagram the incident ray, the reflected ray, the normal at the point of incidence and the angles of incidence and reflection.

(ii) A witness giving evidence in a court case said that the time the crime was committed was at 6.20. The detective knew that this must be wrong and realised that the witness had seen the clock in a plane mirror. What was the time of the crime?

Question 8.3

(a) Explain why light and heat from the sun disappear simultaneously when the sun becomes totally eclipsed.

(b) Draw a diagram illustrating how two glass prisms may be used to make a simple periscope.

(c) Explain why a pond looks shallower than it really is.

Question 8.4

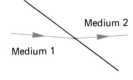

The diagram shows a ray of light crossing an interface between two different media. Which of the following statements is/are correct?

A Medium 1 could be water and medium 2 could be air.

B The frequency of the light in medium 1 is greater than the frequency of the light in medium 2

C The velocity of the light in medium 1 is greater than the velocity of the light in medium 2

D The wavelength of the light in medium 2 is greater than the wavelength of the light in medium 1

Question 8.5

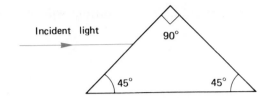

(a) The diagram shows a beam of light incident on a glass prism. Complete the diagram to show the subsequent path of the light until it emerges from the prism (the critical angle for glass is 42°).

(b) How would you use the same prism to produce a spectrum? Draw a diagram to illustrate your answer and mark clearly on the diagram the position of the emergent violet light and red light.

Question 8.6

The diagram shows a lens with an object marked O_1 on the principal axis of the lens. Rays from O_1 pass through the lens and converge at its image point I_1. The lens is midway between O_1 and I_1; hence, O_1 and I_1 are each two focal lengths from the lens. Draw rays from O_2 and O_3 in the diagram, showing where they would converge approximately after passing through the lens. Label these image points I_2 and I_3, respectively.

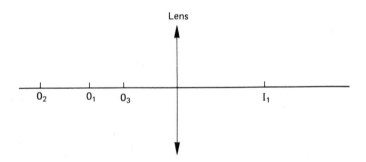

Question 8.7

(a) Draw *two* rays on the figure below from the top of the object O to the top of its image. The image plane is marked with a straight line.

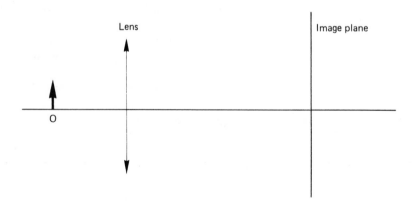

(b) What sort of lens is being used here?

(c) Describe the image by choosing words from *real, vertical, upright, inverted, magnified, diminished*.

(d) A photograph is taken with a lens camera at a shutter speed of $\frac{1}{100}$ s. The area of the aperture is doubled. What shutter speed is now needed to get the same activation of the film?

8.13 Answers and Hints on Solutions to Questions

1. If you are stuck draw a diagram like the one in Example 8.3. A ray from 1 m below his eyes is reflected from the bottom of the mirror to his eyes. A ray from lower down striking the mirror will pass above his eyes.
   Answer **D**

2. (i) See Section 8.1, Fig. 8.1.
   (ii) 5.40 (set a clock to 6.20 and hold it in front of a mirror if you find this question difficult).

3. (a) Both are electromagnetic waves which travel with the same velocity.
   (b) See Example 8.10.
   (c) See Example 8.6.

4. The ray is passing into an optically denser medium in which its velocity and wavelength are less. The ray is bent towards the normal because of the decrease in velocity.
   Answer **C**

5. (a) The light is bent towards the normal on entering the block and is incident on the far glass surface at an angle of incidence which is greater than the critical angle; at this boundary it is totally internally reflected. It hits the bottom face of the prism and is refracted away from the normal as it leaves the prism.
   (b) See Example 8.9.

6. The rays leaving O_2 pass through a point on the principal axis closer to the lens than I_1. Those leaving O_3 pass through a point on the principal axis further from the lens than I_1. I_2 must be at a distance from the lens of more than one focal length.

7. (a) The tip of the image will be where the ray from the top of the object hits the image plane after passing straight through the centre of the lens. *Any* other ray from the top of the object will pass through the tip of the image.
   (b) A converging lens.
   (c) Real, inverted, magnified.
   (d) $\frac{1}{200}$ s (twice the light; hence, half the exposure time).

9 Wave Motion, Sound and Interference

9.1 Wave Motion and Sound

(a) Energy, Speed, Frequency and Wavelength

In all wave motion, a disturbance (energy) travels through a medium without the medium moving bodily with it. The wave transmits energy from the source to the receiver.

The amplitude of the waves is their biggest displacement from the undisturbed level (see Example 9.1), and is a measure of the energy of the waves. In light a greater amplitude results in greater brightness, and in sound a greater amplitude results in a louder note. The frequency is the number of oscillations occurring every second (unit: hertz (Hz)). The wavelength is the distance between successive crests or successive troughs (see Example 9.1).

The speed, v, of any wave is related to the frequency, f, and the wavelength, λ, by the equation $v = f\lambda$. The speed of a wave depends on the medium in which it is travelling. All electromagnetic waves (e.g. radio, light and X-rays) travel at the same speed in a vacuum — namely, 3×10^8 m/s.

(b) Longitudinal and Transverse Waves

In a transverse wave, oscillations are perpendicular to the direction of propagation of the wave (water waves and electromagnetic waves are transverse). In a longitudinal wave the medium oscillates along the direction in which the wave travels (sound waves are longitudinal).

(c) Reflection, Refraction and Interference

Waves can be reflected, refracted, diffracted and show interference effects. Refraction results from a change in speed as the wave crosses a boundary between two different media; the wavelength of the wave also changes, but the frequency does not change (see Example 9.3).

(d) Sound

Every source of sound has some part which is vibrating. Sound needs a medium in which to travel and cannot travel through a vacuum. Sound waves consist of a series of alternate compressions and rarefactions travelling away from the source. The pitch of a note depends on its frequency; if the frequency increases the pitch of the note goes up. The harder you strike a drum, the greater is the amplitude of vibration and the louder is the note heard.

9.2 Diffraction and Interference

If a plane wave passes through a small gap, it spreads out round the edges of the gap. The effect is known as diffraction (see Example 9.8).

Young's slits was a classic experiment which demonstrated the constructive and destructive interference of light from a double slit. It was an important experiment in the establishment of the wave theory of light.

A similar experiment with sound waves uses two loudspeakers connected to the same signal generator and placed about 50 cm apart. The variation in intensity resulting from the interference of the waves from the two sources may be heard by walking along a line parallel to the line joining the two speakers. There are some places where a loud note is heard. These are where compressions from the two sources arrive together, and rarefactions arrive together. Where a compression arrives with a rarefaction, no sound is heard (see Example 9.8).

9.3 Worked Examples

Example 9.1

(a) Draw a labelled diagram to illustrate the meaning of the words *wavelength* and *amplitude*. **(3 marks)**

(b) What is meant by frequency? **(2 marks)**

(c) Water waves are made by a dipper moving up and down 3 times every second. If the velocity of the waves is 12 cm/s, what is the wavelength of the waves? **(3 marks)**

Solution 9.1

(a)

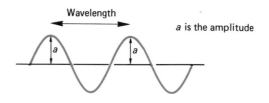

(b) Frequency is the number of complete oscillations (or cycles) in one second.

(c) $v = f\lambda$ [see Section 9.1]

$$12 \text{ cm/s} = 3 \text{ Hz} \times \lambda$$

$$\lambda = \frac{12}{3} \text{ cm} = 4 \text{ cm}$$

Example 9.2

(a) The diagram shows waves spreading out from a point source O and travelling towards a plane reflecting surface. Complete the diagram, showing what happens to the waves as they arrive at, and leave, the reflecting surface. Show the position of the image of the source on your diagram. **(5 marks)**

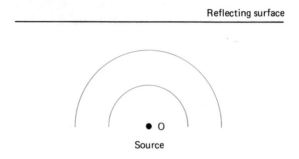

Reflecting surface

● O

Source

(b) Plane waves in a ripple tank have a frequency of 6 Hz. If the wave crests are 1.5 cm apart, what is the speed of the waves across the tank? **(3 marks)**

Solution 9.2

(a)

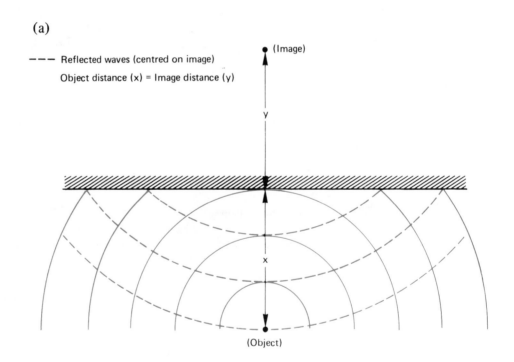

‑ ‑ ‑ Reflected waves (centred on image)

Object distance (x) = Image distance (y)

● (Image)

y

x

● (Object)

(b) $v = f\lambda$ [see Section 9.1]
$v = 6 \text{ Hz} \times 0.015 \text{ m} = 0.09 \text{ m/s}$

Example 9.3

(a) Draw two diagrams showing plane waves crossing a straight boundary and passing into a medium in which their speed is greater, (i) when the wavefronts of the incident wave are parallel to the boundary and (ii) when the wavefronts make an angle with the boundary. **(5 marks)**

(b) What change occurs in (i) the frequency, (ii) the speed and (iii) the wavelength, as a result of refraction when light passes into an optically less dense medium. **(3 marks)**

110

Solution 9.3

(a)

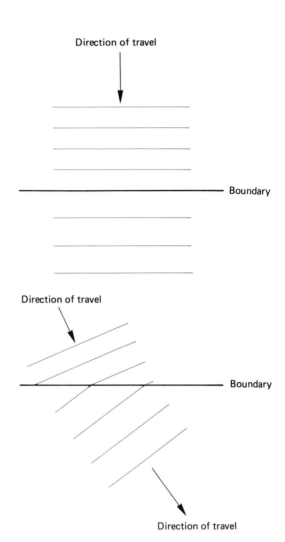

(b) The frequency remains constant. The speed and wavelength increase.

Example 9.4

(a) Describe an experiment to measure the speed of sound in air. **(6 marks)**
(b) A girl observes a man hammering a post into the ground repeatedly and she hears the sound at the same time as he strikes each blow. Explain this observation. If the interval between the blows is 2 seconds and the speed of sound in air is 330 m/s, how far is she from the man? **(4 marks)**

Solution 9.4

(a) Stand 100 metres (measured with a tape measure) from a large vertical wall. Make sure a wall is chosen so that there are no other walls which would produce a substantial echo. One person claps hands at a steady rate and the rate is adjusted until the echo returns at the same time as the next clap. When this is achieved a second person times 50 claps. The sound has travelled 200 m in the time interval between claps. If 50 claps take t seconds then

$$\text{speed of sound} = \frac{200 \text{ m}}{t/50 \text{ s}} = \frac{10\,000}{t} \text{ m/s}$$

(b) She hears the sound from the first blow at the time she sees the second blow.

Distance from man = 330 × 2 = 660 m

She could also be 1320 m from the man if she heard the first blow when she observed the third blow, or 1980 m if she heard the first blow when she observed the fourth blow.

Example 9.5

The table gives the speed of water waves in metres per second for various wavelengths and for four different depths of water.

	Wavelength/m					
	0.001	0.01	0.1	1	10	100
Depth/m 0.1	0.67	0.25	0.40	0.93	0.99	0.99
1	0.67	0.25	0.40	1.25	2.95	3.13
10	0.67	0.25	0.40	1.25	3.95	9.33
100	0.67	0.25	0.40	1.25	3.95	12.5

1. What is the speed of a wave of wavelength 0.1 m at a depth of 10 m?   (1 mark)
2. For a wavelength of 0.1 m or less, what is the relationship between speed and depth?   (2 marks)
3. Describe, in general terms, how the speed varies with the depth, for a wavelength of 100 m.   (2 marks)
4. Describe, in general terms, how the speed varies with the wavelength, for a depth of 100 m.   (2 marks)
5. What is the frequency of a wave of wavelength 10 m at a depth of 10 m?   (3 marks)

Solution 9.5

1. 0.40 m/s.
2. The speed is independent of the depth. [For example, at wavelength 0.1 m at every depth the speed is 0.4 m/s.]
3. As the depth increases, the speed increases.
4. The speed increases as the wavelength increases, increasing very rapidly at large wavelengths.
5. $v = f\lambda$ [see Section 9.1].
   [At a wavelength of 10 m and a depth of 10 m the speed is 3.95 m/s.]
   3.95 m/s = f × 10 m
   $$f = \frac{3.95 \text{ m/s}}{10 \text{ m}} = 0.395 \text{ Hz}$$

Example 9.6

(a) Describe an experiment to show that sound cannot travel in a vacuum.   (6 marks)
(b) Why is the Moon sometimes referred to as 'the silent planet'?   (2 marks)
(c) A sound is sent out from the hull of a ship and the reflection from the bottom of the ocean returns 1 s later. How deep is the ocean? The velocity of sound in water is 1500 m/s.   (4 marks)

Solution 9.6

(a) An electric trembler bell is suspended inside a bell jar from which the air can be removed. When the lowest pressure has been reached, the ringing can no longer be heard (the faint vibration which is audible results from the passage of the sound along the suspension).

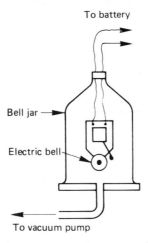

To battery

Bell jar →

Electric bell

To vacuum pump

(b) The Moon has no atmosphere and there is therefore no gas through which the sound can travel. The only way sound can travel is by vibration on the surface.
(c) The sound travels 1500 m in 1 s. This is the time taken for the sound to go to the bottom of the ocean and back to the ship. Depth of ocean = 750 m.

Example 9.7

(a) A microphone is connected, via an amplifier, to a cathode ray oscilloscope. Three different sounds are made, one after the other, near the microphone. The traces produced on the screen are shown in the diagrams below. During the experiment the controls of the oscilloscope were not altered.
  (i) Which trace results from the sound having the highest frequency? (2 marks)
  (ii) Which trace results from the loudest sound? (2 marks)

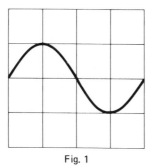
Fig. 1

Fig. 2

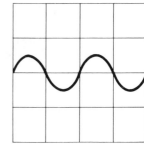
Fig. 3

(b) If in Fig. 1 the time base on the oscilloscope is set at 1 ms/cm and the lines on the grid on the screen are 1 cm apart, what is the frequency of the note which emits the sound giving the trace shown in Fig. 1? (4 marks)

113

Solution 9.7

(a) (i) Fig. 3   [In Fig. 1 and Fig. 2 a whole cycle takes four grid squares. In Fig. 3 a whole cycle only takes two grid squares.]

   (ii) Fig. 2   [The amplitude is greatest in Fig. 2.]

(b) $\left[\text{There is one complete cycle on the screen. } 4 \text{ ms} = \dfrac{4}{1000} \text{ s} = \dfrac{1}{250} \text{ s}\right]$

The time for 1 cycle is 4 ms.

∴ It takes $\dfrac{1}{250}$ s for 1 cycle

∴ There are 250 cycles every second

Frequency = 250 Hz

Example 9.8

(a) Draw diagrams to illustrate what happens when plane waves are incident on a slit, (i) when the width of the slit is large compared with the wavelength of the waves   **(3 marks)**
(ii) when the width of the slit is small compared with the wavelength of the waves

(3 marks)

(b) A student set up a demonstration using two loudspeakers connected to the same oscillator, which was producing a note of fixed frequency. The loudspeakers were placed at A and B and they emitted waves which were in phase. An observer walked along the line PQRS. A loud note was heard at Q and a faint note at R.

(i) On the same axes sketch two graphs showing how the displacement of the vibrating air molecules varies with the time for the disturbance at Q, one for the waves from A and one for the waves from B. On the same axes sketch a third graph showing the displacement at Q for both sets of waves arriving together.   **(4 marks)**
(ii) Repeat (i) for waves arriving at R. What is the relationship between the distances AR and BR?   **(5 marks)**

Solution 9.8

(a)

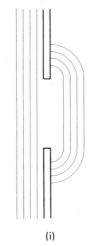 

          (i)   (ii)

(b)

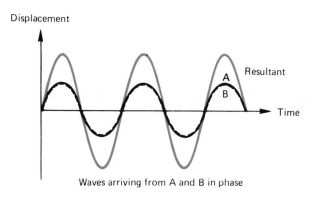

Waves arriving from A and B in phase

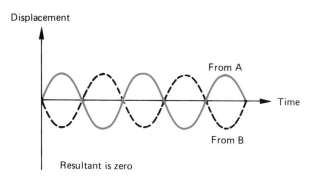

Resultant is zero

$$AR - BR = \lambda/2$$

[When the path difference is $\lambda/2$ then the waves arrive at R 180° out of phase.]

9.4 Have You Mastered the Basics?

1. Can you explain the difference between longitudinal and transverse waves?
2. Can you explain the terms *wavelength, frequency* and *amplitude*?
3. Can you state and use the relationship between velocity, frequency and wavelength of a wave?
4. Do you understand the relationship between pitch and frequency, loudness and amplitude?
5. Can you describe an experiment to determine the velocity of sound in air?
6. A radio programme is broadcast on a wavelength of 1500 m. If the speed of radio waves is 3×10^8 m/s, what is the frequency of the transmission?
7. Can you draw wave diagrams to illustrate reflection and refraction of waves?
8. Can you describe an experiment to illustrate interference of waves?

9.5 Answers and Hints on Solutions to 'Have You Mastered the Basics?'

1. See Section 9.1.
2. See Section 9.1 and Example 9.1.
3. See Section 9.1 and Example 9.2.
4. See Section 9.1(d) and Section 9.1(a).
5. See Example 9.4.
6. $v = f\lambda$ [see Section 9.1] $\Rightarrow 3 \times 10^8 \; \frac{\text{m}}{\text{s}} = f \times 1500 \; \text{m} \Rightarrow f = 200 \; \text{kHz}.$

7. See Examples 9.2 and 9.3.

8. See Section 9.2. Also Example 9.8.

9.6 Questions

(Answers and hints to solutions will be found in Section 9.7.)

Question 9.1

Which one of the following diagrams best illustrates the reflection of a wavefront by a plane reflecting surface in a ripple tank?

————— incident wavefront

————— reflected wavefront   (AEB)

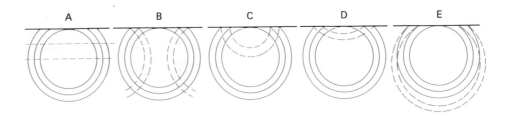

Question 9.2

Figure 1 is a full-size diagram representing the crests of circular water ripples which are travelling outwards and about to meet a plane reflector.

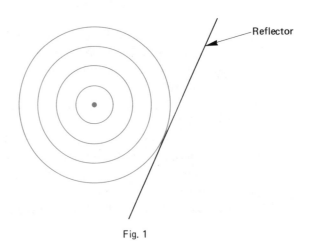

Fig. 1

(a) Describe the source and its action by which a continuous series of such ripples could be produced.   **(3 marks)**

(b) Find from the diagram the wavelength of the ripples.   **(1 mark)**

(c) Add to the diagram the crest of the previous ripple in the series, including its reflected part.   **(2 marks)**

(d) Mark on the diagram the point C on which the reflected ripples are centred.   **(2 marks)**

(e) What happens to the amplitudes of the ripples as they move further from the source? Explain.   **(2 marks)**

(OLE)

Question 9.3

Assuming you have a ripple tank with a suitable quantity of water in it, describe how you would demonstrate (i) refraction and (ii) diffraction of water waves. Your answer should include diagrams showing the result you would expect to observe. **(10 marks)**

Question 9.4

(a) Mention three observations you could make which would help you convince someone that the speed of a wave depends on the medium in which it is travelling. **(3 marks)**

(b) A sound wave has a *wavelength* of 3 m and a *frequency* of 110 Hz. Explain the meaning of the italicised words and calculate the velocity of the wave motion. **(7 marks)**

Question 9.5

(a) Describe an experiment to demonstrate the diffraction of waves. **(4 marks)**

(b) Two loudspeakers are set up about 40 cm apart and each speaker emits the same note.

(i) What would you tell someone to do if you wanted them to hear the interference effects of the two waves? **(2 marks)**

(ii) If the wavelength of the waves is λ, and you are standing at a point P which is at distance S_1P from source S_1, and distance S_2P from source S_2, what is one possible relationship between these distances and λ if you only hear a very faint note? Explain your answer. **(4 marks)**

(iii) At such a point you would expect no sound at all. Why is a faint note usually heard? **(2 marks)**

9.7 Answers and Hints on Solutions to Questions

1. **D**

2. (a) A small sphere being made to bob up and down in the water with a fixed frequency.

   (b) Wavelength = 0.5 cm.

   (c) and (d) See Example 9.2.

   (e) The amplitude decreases as the energy becomes more spread out.

3. Plane waves may be produced by suspending a bar using rubber bands so that the bar is just touching the surface of the water. When the bar is made to vibrate up and down, plane waves are produced. To demonstrate refraction, a shallow area is required. This may be produced by putting a thick sheet of glass on the bottom of the tank at an angle to the oncoming waves. Diffraction will need two vertical barriers arranged with a small gap between them. For diagrams see Examples 9.3 and 9.8.

4. (a) You could mention (i) refraction of light, (ii) the sound of an approaching train heard via the line and through the air, and (iii) a pulse travelling down a rubber tube and a steel spring.

   (b) See Section 9.1 and Example 9.1. Use $v = f\lambda$. Velocity is 330 m/s.

5. (a) See Example 9.8.

   (b) (i) See Example 9.8; (ii) $S_2P - S_1P = \dfrac{\lambda}{2}$. If one source is $\dfrac{\lambda}{2}$ further from you than the other source, then a compression from one source will arrive with a rarefaction from the other source. There will be no resultant disturbance. (iii) The faint note is heard because of reflection from the walls of the room.

10 Circuits, Electrical Units, Household Electricity and Electronics

10.1 Circuits

(a) Basic Units

When an ammeter reads 1 ampere (A), then 1 coulomb (C) of charge is flowing every second. $1\text{ A} = 1\ \dfrac{\text{C}}{\text{s}}$. If a current I passes for a time t, then the charge, Q, which flows is given by $Q = It$.

The potential difference (in volts) between two points is the work done in joules in moving 1 coulomb of charge between them:

$$\text{potential difference} = \frac{\text{work done}}{\text{charge moved}}$$

$$1\text{ V} = 1\ \frac{\text{J}}{\text{C}}$$

The e.m.f. (electromotive force) is the total energy supplied by a source to each coulomb of charge that passes through it, including any energy that may be lost as heat in the source itself as a result of the source's internal resistance. It is measured in $\dfrac{\text{J}}{\text{C}}$ – i.e. volt.

Both e.m.f. and potential difference are sometimes referred to as *voltage*.

The potential difference in volts across the terminals of a cell or generator is the energy which is delivered to the external circuit by each coulomb of charge.

$$\text{Resistance (ohms)} = \frac{\text{potential difference across the object (volts)}}{\text{current passing through the object (amperes)}}$$

or $R = \dfrac{V}{I}$

You must learn to write the above equation as $V = IR$ and $I = V/R$.

An object has a resistance of 1 ohm if a potential difference across it of 1 volt results in a current of 1 ampere passing through it. Most metal conductors have a fixed resistance if their temperature is kept constant. Doubling the length of a conductor doubles its resistance; doubling the cross-sectional area halves its resistance.

If the total e.m.f. of a circuit is E volts, then

$E = I \times$ (total resistance of circuit)

(b) Ohm's Law

Ohm's law states that the current in a conductor is proportional to the potential difference across it provided the temperature is kept constant.

If a conductor obeys Ohm's law, a graph of potential difference against current is a straight line through the origin. The gradient of the graph is the resistance of the conductor (see Example 10.1).

(c) Laws for Series Circuits

(i) The same current passes through each part of the circuit.
(ii) The applied potential difference is equal to the sum of the potential difference across the separate resistors: $V = V_1 + V_2 + V_3$ (Fig. 10.1).
(iii) The total resistance is equal to the sum of the separate resistances: $R = R_1 + R_2 + R_3$.

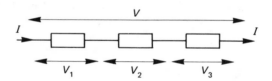

Figure 10.1 Resistors in series. $V = V_1 + V_2 + V_3$.

(d) Laws for Parallel Circuits

(i) The potential difference across each resistor is the same.
(ii) The total current is equal to the sum of the currents in the separate resistors: $I = I_1 + I_2 + I_3$ (Fig. 10.2).

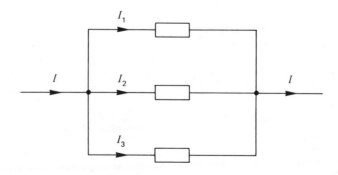

Figure 10.2 Resistors in parallel. $I = I_1 + I_2 + I_3$.

119

(iii) The combined, or total, resistance of a number of resistors in parallel is less than the value of any of the separate resistors and is given by

$$\frac{1}{R} = \frac{1}{R_1} + \frac{1}{R_2} + \frac{1}{R_3}$$

For *two* resistors in parallel

$$\frac{1}{R} = \frac{1}{R_1} + \frac{1}{R_2} \quad \text{or} \quad R = \frac{R_1 \times R_2}{R_1 + R_2}, \quad \text{i.e.} \quad R = \frac{\text{product}}{\text{sum}}$$

(e) Potential divider

See Fig. 10.3. If a p.d. of V_{in} is applied across two resistors R_1 and R_2 in series,

then current in R_1 and $R_2 = \dfrac{V_{in}}{R_1 + R_2}$

p.d. across $R_2 = V_{out} = \left(\dfrac{V_{in}}{R_1 + R_2}\right) \times R_2$

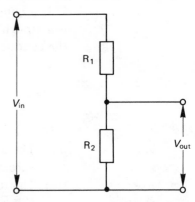

Figure 10.3 (*above right*) A potential divider. If the resistance of R_2 is increased, then V_{out} increases. If the resistance of R_1 is twice the resistance of R_2, then the p.d. across R_1 is twice the p.d. across R_2.

(f) Ring Main and Lighting Circuits

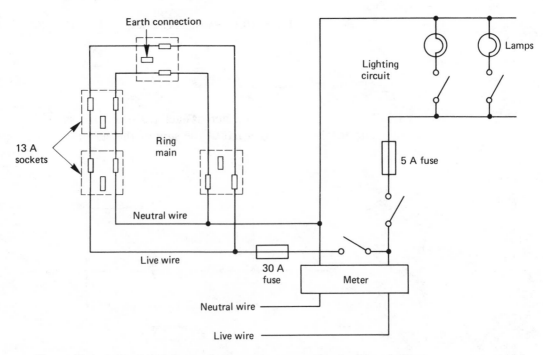

Figure 10.4 In both lighting and ring main power circuits, the appliances are connected in parallel across the supply.

120

The mains supply is delivered to a house by means of two wires called the live and the neutral (see Fig. 10.4). Every appliance is connected in parallel with the supply. Lighting circuits usually have a 5 A fuse. In the ring main, the live, the neutral and the earth wires run in a complete ring round the house.

(g) Conductors, Semiconductors and Insulators

An insulator (e.g. rubber or plastic) is a substance which will not conduct electricity. Conductors (e.g. copper) are good conductors of electricity. In between are the semiconductors (e.g. germanium), which are neither good conductors nor good insulators.

10.2 Electrical Power and Energy

Power (watts) = potential difference (volts) × current (amperes)

or power P $= (V \times I)$ joule/second

Substituting, using the definition of resistance given above, namely

potential difference (V) = current (I) × resistance (R)

we have

power = $(I^2 \times R)$ joule/second

Thus, the energy dissipated in a resistor is $I^2 \times R$ joule/second. Hence, the total energy dissipated in t seconds is given by

energy dissipated = $I^2 \times R \times t$ joule

A meter supplied by the electricity company records the energy consumption in kilowatt-hours. One kilowatt-hour (kWh) is the total amount of energy supplied to a 1 kW appliance when it is connected to the supply for 1 h. As a 1 kW appliance is supplied with energy at a rate of 1000 J/s, a total of 3.6×10^6 J of energy will be supplied in 1 h (3600 s); thus

1 kWh = 3.6×10^6 joule.

The cost of running an electrical appliance may be calculated from the equation

cost = (number of kilowatts) × (hours) × (cost of a kilowatt-hour)

10.3 Earthing and Wiring

Appliances with metal casing have an earthed wire connected to the metal casing. This is a safety device. For example, if the element in an electric fire breaks and the live end touches the metal case, the connection to earth has a very low resistance, and the current surges to a large value. The fuse in the circuit is designed to melt and break the circuit when the current reaches a predetermined value. The large current that results when the live wire touches the case melts the fuse, the circuit is broken and is thus safe. In many modern circuits the fuse is replaced by a circuit-breaker. 'Double insulated' appliances have their metal parts surrounded by thick plastic, making it impossible for the user to touch a metal part. When wiring double insulated appliances, no earth lead is necessary. Damp surroundings can be dangerous as moisture greatly reduces the resistance of the skin. In a three-pin plug, the earth wire is yellow and green, the live wire is brown and the neutral wire is blue (see Question 10.7).

An earth leakage circuit-breaker (also called an earth leakage trip) is a main switch automatically operated by a small current passing (leaking) to earth. An important characteristic of the earth leakage circuit-breaker is that only a very small earth current is needed to operate it. It is easily reset and, unlike a fuse, no replacement is needed. Earth leakage circuit-breakers should always be used as a safety precaution when electrical equipment is used out of doors, especially in damp conditions.

10.4 Electronics

(a) Light-dependent Resistors (LDRs), Light-emitting Diodes (LEDs) and Thermistors

LDRs, LEDs and thermistors are examples of *transducers* (i.e. they convert energy from one form to another).

The resistance of a light-dependent resistor decreases as the intensity of the light falling on it increases. (LDRs are used in light meters.)

LEDs are semiconductor diodes which convert electrical energy into light energy. To avoid damage by large currents, a protective resistor is usually connected in series with the LED.

Figure 10.5 shows a circuit with an LDR and an LED in it. When light shines on the LDR, the LED switches off.

The resistance of most thermistors decreases as their temperature increases.

(b) Gates

The symbols and truth tables of a number of gates are shown below.

Name of gate	Symbol	Truth table			Description of gate
AND	A, B — O.P.	A	B	O.P	The output is high if both A *AND* B are high
		0	0	0	
		0	1	0	
		1	0	0	
		1	1	1	
NAND	A, B — O.P.	0	0	1	Opposite of *AND* gate
		0	1	1	
		1	0	1	
		1	1	0	
OR	A, B — O.P.	0	0	0	The output is high if A *OR* B *OR* both are high
		0	1	1	
		1	0	1	
		1	1	1	
NOR	A, B — O.P.	0	0	1	Opposite of *OR* gate. Output high if neither A *NOR* B is high
		0	1	0	
		1	0	0	
		1	1	0	
NOT	A — O.P.	A	O.P		An inverter
		0	1		
		1	0		

The circuit shown in Fig. 10.5 illustrates the use of an LDR and a NOT gate to switch on an LED when it gets dark. If a thermistor replaced the LDR, then the light would come on as the temperature fell.

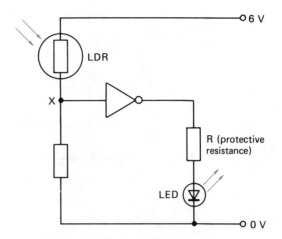

Figure 10.5 When light shines on the LDR, its resistance goes down and the voltage at X goes up. The output from the NOT gate goes from high to low and the LED switches off. The reverse happens when it gets dark.

(c) The Bistable

Two NAND gates may be combined as shown in Fig. 10.6 to form a bistable. Suppose Q is at logic 0 and $\overline{Q}$ is at logic 1, and A and B are both high (logic 1). If A is made low (logic 0), then Q goes high (logic 1). If B is then made low, Q resets to a low. If inputs A and B are both high, the circuit remains in one of its two stable states, depending on which of the AB inputs was last at a low voltage. With

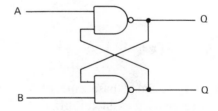

Figure 10.6 Two NAND gates connected together to make a bistable.

slight modifications the bistable may be switched from one state to the other every time a pulse is fed to the input. The bistable forms the basis of memory circuits and, with additional logic, binary counters.

(d) Capacitors

A capacitor may be used to store charge and energy. In the circuit shown in Fig. 10.7 it is used to produce a time delay. When the switch S is closed, the output from the NOT gate is high and the lamp is off. When the switch S is opened, the capacitor charges up through the resistor R. When the voltage at A is high enough, the output from the NOT gate goes low and the lamp lights. Increasing the value of the capacitor or the resistor will increase the time delay.

123

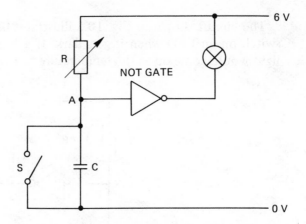

Figure 10.7 A capacitor being used to produce a time delay.

(e) Transistors

A transistor is a semiconductor device. It is considered to be 'off' when very little current passes through the collector circuit and 'on' when a much larger current passes through in the collector circuit. The transistor stays 'off' unless the base voltage and, hence, the base current, rise above a certain minimum value. When the base current rises above this minimum value, a much larger current passes through the collector circuit (Fig. 10.8). A small base current can therefore be used to switch the transistor 'on' and 'off'. See Example 10.20.

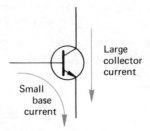

Figure 10.8 A small base current results in a large collector current.

Because a small base current can cause a large collector current, a transistor may also be used as an amplifier.

(f) Relays

A relay is a switch which is operated by an electric current. Often the relay consists of a solenoid with a switch inside it. The switch closes when a current passes through the solenoid. Figure 10.9 shows a relay being used to switch on a large current, using only the current from a battery. One common type of relay is a reed switch (Fig. 10.10) placed in a solenoid.

(g) A Bridge Rectifier Circuit

A semiconductor diode is a device which allows current to pass easily in one direction only. The four diodes shown in Fig. 10.11 are arranged to form a 'bridge rectifier'. They 'steer' the current in the desired direction. The current can only

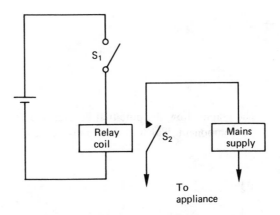

Figure 10.9   When the switch S_1 is closed, a current passes through the relay coil. The switch S_2 closes and the power to the appliance is switched on.

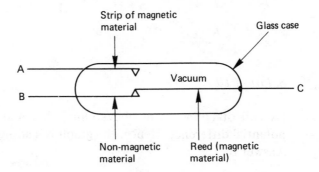

Figure 10.10   A reed switch. In the presence of a magnetic field, the reed is attracted to strip A, and A and C are connected.

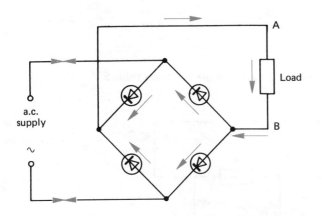

Figure 10.11   A bridge rectifier circuit. The current always flows in the same direction through the load.

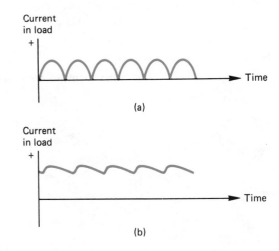

Figure 10.12   (a) Unsmoothed d.c. (b) Smoothed d.c.

pass in the direction shown by the arrows. It will always pass through the load from A to B.

Figure 10.12(a) shows the current through the load. The output from the bridge rectifier is said to be full-wave rectified.

If a capacitor is placed across the load (i.e. across the output from the bridge rectifier) the variations in the current are reduced. The capacitor charges up while the output voltage is rising, and discharges when the output voltage is falling. The smaller variations in the current which remain are known as the ripple current (Fig. 10.12(b)).

10.5 Worked Examples

Example 10.1

The graphs show the potential difference across a component plotted against the current in the component. Which of the graphs would be obtained for a coil of copper wire?

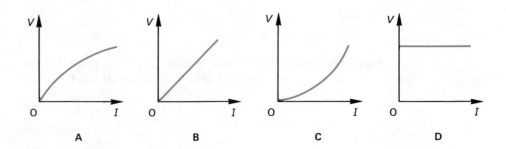

A   B   C   D

Solution 10.1

[A coil of copper wire obeys Ohm's law and the current is proportional to the potential difference. Hence, the graph is a straight line through the origin.]
Answer **B**

Example 10.2

In the circuit shown below, the ammeter reads 2 A and the voltmeter reads 5 V.

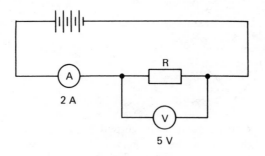

The value of the resistance R is
A 0.4 Ω   B 2.5 Ω   C 3.0 Ω   D 10.0 Ω

Solution 10.2

$$\left[R = \frac{V}{I} \text{ (see Section 10.1)} = \frac{5 \text{ V}}{2 \text{ A}} = 2.5 \text{ Ω} \right]$$

Answer **B**

Example 10.3

(i) The diagram shows a potential divider circuit. The e.m.f. of the cell is 12 V and it has negligible internal resistance.

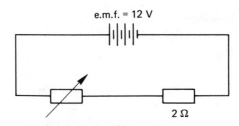

e.m.f. = 12 V

2 Ω

What happens to the potential difference across the 2 Ω resistor as the resistance of the variable resistor is increased?

Explain your answer.

(ii) When the variable resistor is set at 4 Ω, what is
  (a) the current in the 2 Ω resistor?
  (b) the potential difference across the 2 Ω resistor?

Solution 10.3

(i) The potential difference across the 2 Ω resistor decreases. The current passing through it will decrease because the circuit resistance has increased, and hence the p.d. across it, which is given by $V = IR$, will decrease.

(ii) (a) Total resistance of circuit = (2 Ω + 4 Ω) = 6 Ω.

$$\text{Current flowing} = \frac{V}{R} \quad [\text{see Section 10.1}]$$

$$= \frac{12 \text{ V}}{6 \text{ Ω}} = 2 \text{ A}$$

[In a series circuit the current is the same everywhere. The current in every part of the circuit is 2 A.]

(b) Potential difference across 2 Ω is given by

$V = IR$ [see Section 10.1]
$V = 2 \text{ A} \times 2 \text{ Ω} = 4 \text{ V}$

[Notice that the p.d. across the 4 Ω resistor is 8 V and that the ratio between the p.d.'s is equal to the ratio between the resistors.]

Example 10.4

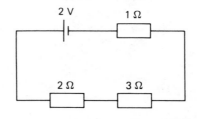

2 V   1 Ω

2 Ω   3 Ω

The potential difference, in V, across the 3 Ω resistor, is

A $\frac{1}{9}$ B $\frac{1}{2}$ C 1   D $\frac{6}{5}$ E 2   (AEB)

Solution 10.4

[The total resistance is 6 Ω. $V = IR$. $\therefore$ 2 V = $I \times 6$ Ω, where I is the current in the circuit. Hence, $I = \frac{1}{3}$ A. For the 3 Ω resistor $V_{3\Omega} = I \times 3$ $\Omega = (\frac{1}{3} \times 3)$ V = 1 V.]

Answer   **C**

Example 10.5

A 12 V battery is used, together with a potential divider, to provide a variable voltage across a 12 V car bulb. Sketch the circuit you would use to measure the current in the lamp for various voltages across it. Sketch the graph you would expect to obtain.   **(8 marks)**

Solution 10.5

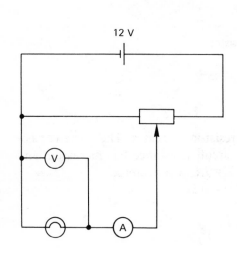
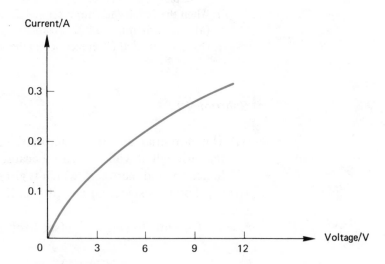

[As the slider is moved from the right-hand end of the variable resistor to the left-hand end, the voltage across the bulb varies from 12 V to 0 V (this arrangement is known as a potential divider). Remember that voltmeters are connected across the appliance and ammeters in series.]

Example 10.6

   (i) A current of 2 A is passing through a circuit. How much charge flows past any point in 5 s?   **(2 marks)**
   (ii) 120 J of work is done when 10 C passes through a bulb. What is the p.d. across the bulb?   **(2 marks)**
   (iii) A 60 W 240 V lamp is connected to a 240 V mains.
      (a) What current does it take?   **(2 marks)**
      (b) What is its resistance?   **(2 marks)**
      (c) How much does it cost to keep it on for 50 hours if the cost of a kWh is 6 p?
          (3 marks)

Solution 10.6

(i) $Q = It$ [see Section 10.1. Remember that 1 A = 1 C/s]

$$Q = \left(2 \ \frac{C}{s}\right)(5 \ s) = 10 \ C$$

128

(ii) p.d. = $\dfrac{\text{work done}}{\text{charge moved}}$ [see Section 10.1. Remember that p.d. is energy per coulomb]

$$= \frac{120 \text{ J}}{10 \text{ C}} = 12 \text{ V}$$

(iii) (a) Power $= (V \times I)$ [see Section 10.2]
60 W $= 240 \; V \times I$

$$I \;\; = \frac{60 \text{ W}}{240 \text{ V}} = 0.25 \text{ A}$$

(b) $R = \dfrac{V}{I}$ [see Section 10.1]

$$R = \frac{240 \text{ V}}{0.25 \text{ A}} = 960 \; \Omega$$

(c) Cost = (number of kilowatts) × (hours) × (cost of kilowatt-hour) [see Section 10.2]

$$= \left(\frac{60}{1000} \text{ kW} \right) (50 \text{ h}) \left(6 \; \frac{\text{P}}{\text{kWh}} \right) = 18 \text{ p}$$

Example 10.7

(i) Why is a three-pin mains plug fitted with a fuse?   (2 marks)
(ii) What fuse would you put in a plug if the plug is connected to an appliance labelled 750 W 250 V?   (3 marks)

Solution 10.7

(i) The fuse is a safety device. If the current exceeds a certain value, the fuse will melt, breaking the circuit and disconnecting the appliance from the supply.
(ii) Power $= (V \times I)$ [see Section 10.2]
750 W $= 250 \; V \times I$

$$I \; = \frac{750 \text{ W}}{250 \text{ V}} = 3 \text{ A}$$

A 5 A fuse would be fitted. [Fuses fitted to household appliances are usually 5 A, 10 A or 13 A.]

Example 10.8

Three resistors, each of resistance 4 Ω, are to be used to make a 6 Ω combination. Which arrangement will achieve this?

(L)

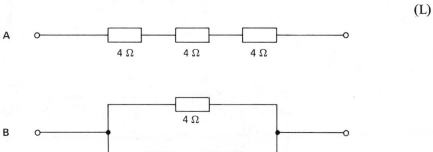

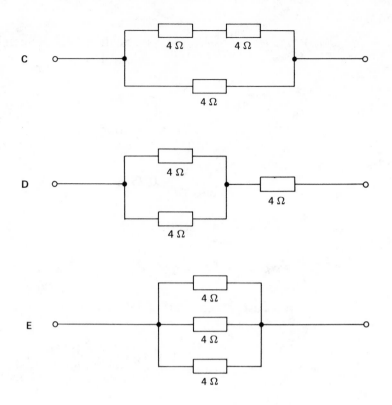

Solution 10.8

[When answering this type of question, don't waste time calculating the effective resistance of every combination. Glance at the various answers and see if you can spot the correct one, then check it by calculation. In this case a glance at **D** will tell you that this is the correct answer. The two 4 Ωs in parallel have an effective resistance of 2 Ω. This 2 Ω is connected in series with 4 Ω and hence the total resistance is 6 Ω].

Answer **D**

Example 10.9

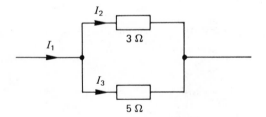

The diagram shows two resistors connected in parallel. Which of the following statements is correct?

A $I_1 = I_2 - I_3$

B $I_1 = I_3 - I_2$

C $5I_2 = 3I_3$

D $3I_2 = 5I_3$

Solution 10.9

[In a parallel circuit the potential difference across each resistor is the same; therefor $I_2 \times 3\ \Omega = I_3 \times 5\ \Omega$. The current divides in the same ratio as that between the resistances, the larger current passing through the smaller resistance.]
Answer **D**

Example 10.10

A 3 V battery of negligible internal resistance is connected in series with a 3 Ω resistor and an ammeter as shown in the diagram.

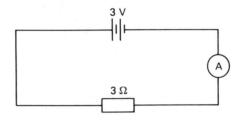

(a) What does the ammeter read? (2 marks)
(b) A 6 Ω resistor is connected across the 3 Ω resistor. What is the new reading on the ammeter? (4 marks)

Solution 10.10

(a) $I = \dfrac{V}{R} = \dfrac{3\ \text{V}}{3\ \Omega} = 1\ \text{A}$

(b) For resistances in parallel,

$$R = \frac{\text{product}}{\text{sum}} \; [\text{see Section 10.1(d)}] \; = \frac{3 \times 6}{3 + 6}\,\Omega = \frac{18}{9}\,\Omega = 2\Omega$$

$$I = \frac{V}{R} = \frac{3\ \text{V}}{2\ \Omega} = 1.5\ \text{A}$$

[Notice that when a resistance is added in parallel, the resistance of the circuit decreases and the current increases.]

Example 10.11

(a) Define resistance.
(b) The diagram shows an incomplete circuit for an experiment to investigate how the resistance of a torch bulb varies with the current flowing through it.
(i) Add to the circuit diagram an ammeter for measuring the current through the bulb and a voltmeter for measuring the p.d. across the bulb.

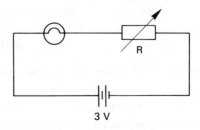

(ii) State clearly how you would obtain the readings needed to carry out the investigation.

(iii) How would you calculate the resistance of the bulb?

(iv) If the bulb is 2.5 V and takes 0.25 A at its working temperature, calculate the resistance of the bulb at the working temperature.

(v) The resistance of the bulb when the filament is cold is 5 Ω. Sketch the graph you would expect to obtain if you plotted resistance against current for the bulb.

(c) The diagram shows an electrical circuit containing a battery of e.m.f. 3 V, an ammeter of negligible internal resistance and three resistors with resistances shown.

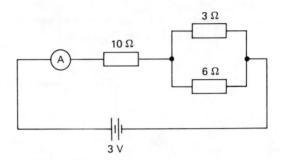

What is (i) the resistance of the parallel combination, (ii) the reading on the ammeter and (iii) the potential difference across the 3 Ω resistor?

Solution 10.11

(a) Resistance is defined by the equation

$$\text{resistance (in ohms)} = \frac{\text{potential difference across object (in volts)}}{\text{current flowing through object (in amperes)}}$$

(b) (i)

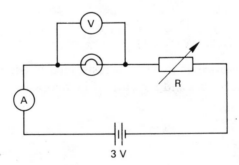

(ii) The resistance R is set at its highest value and the readings on the ammeter and the voltmeter are recorded. The value of the resistance R is decreased and the reading on each meter again recorded. Continuing in this way a series of readings are obtained. The readings are continued until the bulb is burning brightly.

(iii) The resistance for each pair of readings is calculated from the definition given in (a), that is by dividing the voltmeter reading by the ammeter reading.

(iv) $R = \dfrac{V}{I} = \dfrac{2.5 \text{ V}}{0.25 \text{ A}} = 10 \ \Omega$

(v)

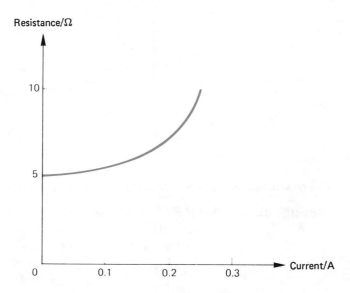

[The same circuit is used to investigate the voltage/current relationship for other components such as a resistor or a diode.]

(c) (i) $R = \dfrac{\text{product}}{\text{sum}}$ [see Section 10.1(d)] $= \dfrac{3 \times 6}{3 + 6}\ \Omega = \dfrac{18}{9}\ \Omega = 2\ \Omega$

(ii) $I = \dfrac{V}{R} = \dfrac{\text{voltage of cell}}{\text{total resistance}} = \dfrac{3\ \text{V}}{12\ \Omega} = 0.25\ \text{A}$

(iii) p.d. across 3 Ω and 6 $\Omega = IR = 0.25\ \text{A} \times 2\ \Omega = 0.5\ \text{V}$

Example 10.12

Define *potential difference*. **(2 marks)**

A coil of resistance 20 Ω is joined in series with a coil X and a D.C. source of e.m.f. 15 V. If the potential difference across the 20 Ω coil is to be 10 V, calculate the resistance of X. (Neglect the resistance of the source and connecting leads.) Draw the circuit diagram; include an ammeter and a voltmeter to check this value for the resistance of X. What readings would you expect on these meters. **(6 marks)**

Calculate the energy dissipated in the 20 Ω coil in 10 minutes if the current remains steady. **(3 marks)**

If X and the 20 Ω coil had been joined in parallel, what would the current from the source have been? **(4 marks)**

(SUJB)

Solution 10.12

The potential difference between two points is the work done in joules in moving 1 coulomb of charge between them.

p.d. across 20 Ω coil is 10 V; therefore current through circuit $= \dfrac{10}{20} = 0.5$ A.

10 V across the 20 Ω coil means that there is 5 V across X.

$R = \dfrac{V}{I} = \dfrac{5\ \text{V}}{0.5\ \text{A}} = 10\ \Omega$. X has a resistance of 10 Ω.

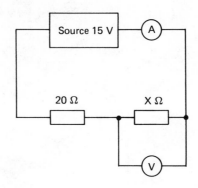

The voltmeter reads 5 V and the ammeter reads 0.5 A.

Energy dissipated $= I^2 Rt$ [see Section 10.2: the energy dissipated is the heat produced]

$$= \{(0.5)^2 \times 20 \times (10 \times 60)\} \text{ J}$$

$$= 3000 \text{ J}$$

For resistances in parallel,

$$R = \frac{\text{product}}{\text{sum}} \text{ [see Section 10.1(d)]} = \frac{20 \times 10}{20 + 10} \Omega = \frac{20}{3} \Omega$$

and

$$I = \frac{V}{R} = 15 \text{ V} \div \frac{20}{3} \Omega = \left(15 \times \frac{3}{20}\right) \text{A} = 2.25 \text{ A}$$

Example 10.13

A boy designs a model railway signal to operate from the circuit shown below.

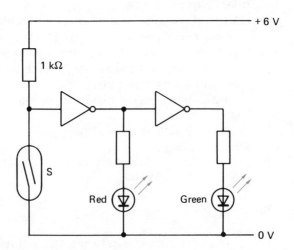

The switch marked S is a reed switch which closes when a magnet is brought close to it. The light-emitting diodes are the lamps for the signal.
  (i) When the switch S is open, which of the two lights are lit? Explain your answer. Why can both lights never be alight at the same time? **(4 marks)**
  (ii) The boy fixes a magnet underneath the train and fixes the reed switch on the track so that the train passes over it. What happens to the signal lights as the train passes over the reed switch? **(4 marks)**

Solution 10.13

(i) Green. When S is open, the input to the first NOT gate is high. The output is low, so the red light is off. The output from the second gate is high, and the green light is on.

(ii) When the train passes over the reed switch, it closes, and the input to the first gate goes low. Its output goes high, turning on the red light. The output from the second gate goes low and the green light goes off.

Example 10.14

In the circuit shown below the switch S is closed when the relay coil is energised.

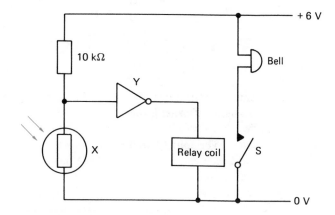

(i) Name the component labelled X. **(1 mark)**

(ii) Complete the truth table for the gate labelled Y.

Input	Output
0	
1	

What is the gate called which is represented by the above truth table? **(2 marks)**

(iii) What happens to the input voltage of the logic gate when light shines on the component X? Explain your answer. **(3 marks)**

(iv) Explain what happens in the rest of the circuit as a result of light shining on the component X. **(3 marks)**

(v) What is the value of the input voltage at the logic gate if the resistance of the component X is 20 kΩ. **(3 marks)**

Solution 10.14

(i) Light-dependent resistor.

(ii)

Input	Output
0	1
1	0

A NOT gate

(iii) It goes low. The resistance of the LDR goes down and the voltage input to the gate goes low.

(iv) The output from the gate goes high and a current passes through the relay coil. This switches on the switch S and the bell rings.

(v) The voltage across X is twice the voltage across the 10 kΩ resistor, so the voltage across the 10 kΩ resistor is 2 V [see Section 10.1(e) and Example 10.3]. The input voltage to the gate is therefore 6 V − 2 V = 4 V.

Example 10.15

(a) Draw a truth table for the two-input NAND gate shown in Fig. 1. **(2 marks)**

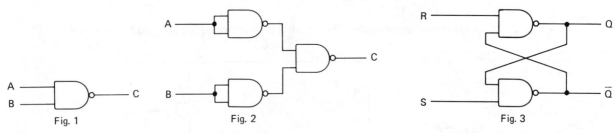

Fig. 1   Fig. 2   Fig. 3

(b) Three NAND gates are connected as shown in Fig. 2. Draw the truth table for the arrangement. What is the resulting gate called? **(3 marks)**

(c) Two NAND gates are connected as shown in Fig. 3. Complete the truth table shown below. The initial condition is shown on the first line of the sequence and you must make each change in the sequence indicated, moving one line at a time down the table. **(4 marks)**

		R	S	Q	$\overline{Q}$
sequence	1	1	1	0	1
	2	0	1		
	3	1	1		
	4	1	0		
	5	1	1		

(d) The above logic is that of the burglar system shown in Fig. 4. The alarm rings when it receives a logic 1 from the output of the NAND gate to which it is connected. A switch will close when a window in the house is opened and open again when the window is closed. Explain why

  (i) the alarm rings when one of the switches is closed, **(3 marks)**

  (ii) the alarm does not stop ringing if the burglar closes the window. **(3 marks)**

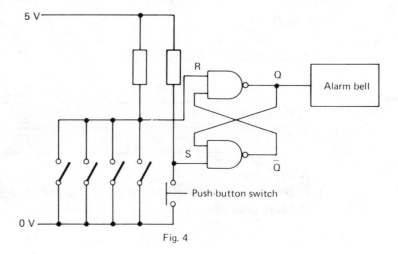

Fig. 4

136

(e) How can the alarm be switched off and reset? **(2 marks)**

(f) Why are two NAND gates connected as shown in Fig. 2 referred to as (i) a *bistable*, (ii) a *flip-flop*? **(3 marks)**

Solution 10.15

(a)

A	B	C
0	0	1
0	1	1
1	0	1
1	1	0

(b)

A	B	C
0	0	0
0	1	1
1	0	1
1	1	1

An OR gate

[The first two NAND gates which have their two inputs joined together are NOT gates and invert the input.]

(c)

	R	S	Q	$\overline{Q}$
1	1	1	0	1
2	0	1	1	0
3	1	1	1	0
4	1	0	0	1
5	1	1	0	1

(d) (i) When one of the switches is closed the input R goes low (logic 0). As shown in the table above, when R goes to logic 0, Q goes to logic 1 and the alarm will ring.

(ii) Closing the window causes the logic of R to change from logic 0 to logic 1 and, as shown in the table above, the output Q remains at logic 1 and the alarm continues ringing. [The alarm is said to be latched.]

(e) Close all the windows, then push and release the push-button switch.

(f) It is called a bistable because it has two stable states. The outputs are either logic 0 and logic 1, or logic 1 and logic 0. It is called a flip-flop because it 'flips' from one state to the other when R goes from logic 1 to logic 0 and 'flops' back again when S goes from logic 1 to 0.

Example 10.16

(a) The basic building block in many electronic circuits is the NAND gate. The truth table for a NAND gate is

Input 1	Input 2	Output
0	0	1
0	1	1
1	0	1
1	1	0

In the circuit shown, five NAND gates labelled g_1, g_2, g_3, g_4 and g_5 are used. Complete the truth table below. A and B are the two inputs; X and Y are the two outputs.

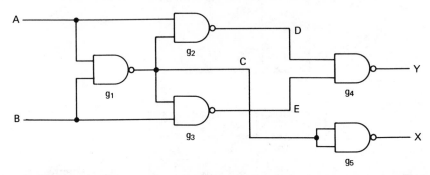

(8 marks)

A	B	C	D	E	X	Y
0	0	1				
0	1	1				
1	0					
1	1					

(b) What is the name of the arithmetic operation which the above circuit performs?

(2 marks)

Solution 10.16

(a)

A	B	C	D	E	X	Y
0	0	1	1	1	0	0
0	1	1	1	0	0	1
1	0	1	0	1	0	1
1	1	0	1	1	1	0

[Complete column C, using the NAND gate truth table. Complete each of the other columns in turn, using the NAND gate truth table – e.g. for gate g_2, A and C are the inputs, and the output is D; for gate g_4 the inputs are D and E, and the output is Y.]

(b) It is a binary (half) adder. [The last two columns, X and Y, are the sum of A and B in binary numbers.]

138

Example 10.17

In the circuit shown below the inputs to the first logic gate are both at logic level '1'. The output from the second gate is also at logic level '1'.

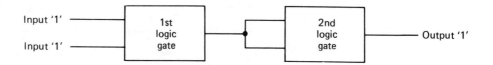

The two gates could be
A the first an AND gate and the second a NAND gate
B the first a NAND gate and the second an AND gate
C the first an AND gate and the second a NOR gate
D both NAND gates

Solution 10.17

[Use the logic tables in Section 10.4(b).]
Answer **D**

Example 10.18

You are required to design a circuit so that an alarm bell rings when a push-button switch is closed (i.e. is taken to logic 1), *and at the same time* it is also daylight *or* raining *or* both. In order to do this, you are provided with a light sensor and a water sensor. The output from the light sensor is high (logic 1) when it is daylight and the output from the water sensor is high (logic 1) when it is raining. The table below summarises the requirements.

Push-button switch	Light sensor	Water sensor	Output
0	0	0	0
0	0	1	0
0	1	0	
0			
1			
1			
1			
1			

(a) Complete the table.   **(5 marks)**
(b) The desired result may be achieved by using two 2-input logic gates. Draw a circuit you could use showing the two logic gates. The details of the bell circuit are not required.
                                                                                          (5 marks)

Solution 10.18

(a)

Push-button switch	Light sensor	Water sensor	Output
0	0	0	0
0	0	1	0
0	1	0	0
0	1	1	0
1	0	0	0
1	0	1	1
1	1	0	1
1	1	1	1

(b)

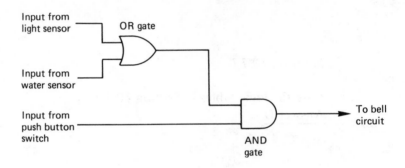

Example 10.19

The circuit drawn below is that used in a tomato waterer. The contacts labelled *A* are buried in the soil.

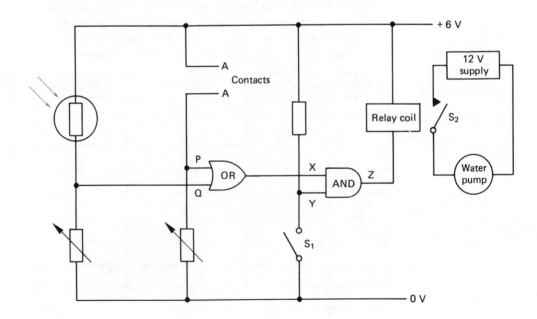

(i) Draw a truth table for the OR gate   **(2 marks)**
(ii) Draw a truth table for the AND gate   **(2 marks)**

(iii) What can you say about the logic states of the inputs X and Y to the AND gate if the water pump is off? Explain your answer. **(4 marks)**

(iv) What happens to the water pump if it is off and the switch S_1 is closed. Explain your answer. **(4 marks)**

(v) Complete the truth table below if S_1 is open.

P	Q	X	Y	Z
0	0			
0	1			
1	0			
1	1			

(4 marks)

(vi) When the soil is very dry, it is dark and the switch S_1 is open, is the water pump on or off? Explain your answer. **(4 marks)**

Solution 10.19

(i)

P	Q	X
0	0	0
0	1	1
1	0	1
1	1	1

(ii)

X	Y	Z
0	0	0
0	1	0
1	0	0
1	1	1

(iii) They must both be high. If X and Y are high, then Z is high, so there is no voltage across the relay coil. No current passes through the coil and S_2 is open.

(iv) The pump starts. When S_1 is closed, the voltage input at Y is low and Z goes low. There is a voltage drop across the relay coil, a current passes and the switch S_2 closes, starting the pump.

(v)

P	Q	X	Y	Z
0	0	0	1	0
0	1	1	1	1
1	0	1	1	1
1	1	1	1	1

(vi) On. When the soil is dry and it is dark, inputs P and Q are both low. As the truth table above shows, in this situation Z is low. There is a voltage drop across the relay coil, a current passes through it, and the switch S_2 closes, turning the pump on.

Example 10.20

In the circuit shown below the lamp L is lit when the light falling on the LDR falls below a certain intensity.

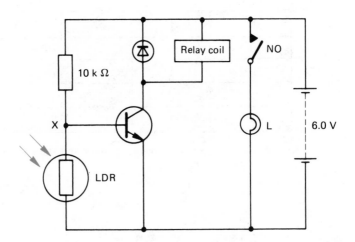

(i) What do the letters 'LDR' stand for? **(2 marks)**

(ii) What do the letters 'NO' stand for? **(2 marks)**

(iii) The resistance of the LDR varies between 300 Ω and 10 kΩ. What is the voltage at the point X when its resistance is
(a) 300 Ω
(b) 10 kΩ? **(6 marks)**

(iv) What change occurs in the current flowing from the collector to the emitter of the transistor as the light falling on the LDR suddenly decreases in intensity so that the resistance of the LDR changes from 300 Ω to 10 kΩ? **(2 marks)**

(v) Explain why this change in current switches the lamp on. **(3 marks)**

Solution 10.20

(i) Light-dependent resistor.

(ii) Normally open.

(iii) Let I be current through the 10 kΩ resistor and then

(a) $V = IR$

$$6 \text{ V} = I\,(10\,000 + 300)\ \Omega$$

$$\therefore I = \frac{6}{10\,300}\ \text{A}$$

Voltage drop across 10 kΩ resistor $= \left(\dfrac{6}{10\,300} \times 10\,000\right) \text{V} = 5.82 \text{ V}$

$V_X = (6.0 - 5.82) \text{ V} = 0.18 \text{ V}$

(b) If the resistance of the LDR is equal to the resistance of the resistor then the voltage drop across each will be the same, namely 3.0 V. Hence $V_X = 3.0$ V.

(iv) The voltage at X changes from 0.18 V to 3 V and the transistor switches on. A current passes from the collector to the emitter.

(v) The current passing from the collector to the emitter flows through the relay. The switch closes and the lamp lights.

Example 10.21

In only one of the circuits below the lamp lights. Which circuit is it?

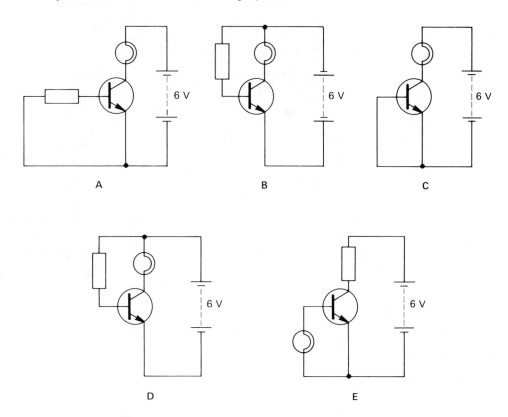

A   B   C

D   E

Solution 10.21

[When a small current passes from the base to the emitter, this switches 'on' the transistor, causing a large current to pass from the collector to the emitter. In B the battery is the wrong way round.]

Answer **D**

Example 10.22

In the circuit below a voltmeter V is connected across lamp L.

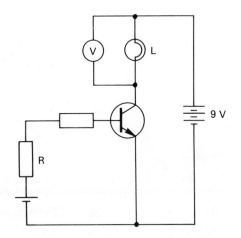

Which of the following changes *could* occur at lamp L and voltmeter V, if the resistor R is reduced in value?

	Lamp *L*	Reading on voltmeter *V*
A	goes out	decreases
B	goes out	increases
C	no change	no change
D	lights up	decreases
E	lights up	increases

(SEB)

Solution 10.22

[If the current in the resistor R is increased the current in the base of the transistor will increase. This could switch the transistor on and the lamp could light. If this happened a current would pass through the lamp and the voltage across it would increase.]

Answer **E**

10.6 Have You Mastered the Basics?

1. Can you explain the meaning of a coulomb, potential difference, a volt, resistance, an ohm, e.m.f. and kilowatt-hour?
2. Can you write and use an equation for calculating the power dissipated in a resistor?
3. Do you understand Ohm's law?
4. Do you understand the laws for series and parallel circuits concerning (i) the total resistance, (ii) the relationship between potential differences and (iii) the relationship between currents?
5. A power supply of e.m.f. 36 V and negligible internal resistance is connected in a circuit as shown in the diagram. What is

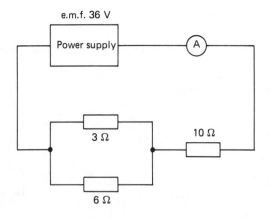

   (i) the current flowing in the ammeter?
  (ii) the current through the 3 Ω resistor?
 (iii) the p.d. across the 10 Ω resistor?
 (iv) the heat energy dissipated every second in the 10 Ω coil?

6. Can you explain how a potential divider works?

7. Can you explain how a house is wired?

8. Can you describe the use of a relay and how it works?

9. How much does it cost to run a 2 kW fire for 3 hours if electricity costs 6 p per kilowatt-hour?

10. Can you draw truth tables for the different logic gates and for a combination of logic gates?

11. Can you explain how a bistable may be used as a latch?

12. Do you understand how LEDs, LDRs, thermistors and capacitors are used in circuits?

13. Can you describe circuits to make (i) a burglar alarm, (ii) light-sensitive and time-delay switches?

14. Can you describe how a transistor may be used as a switch?

10.7 Answers and Hints on Solutions to 'Have You Mastered the Basics?'

1. See Sections 10.1 and 10.2.
2. See Section 10.2 and Example 10.12.
3. See Section 10.1(b) and Examples 10.1, 10.2 and 10.3.
4. See Section 10.1(c), (d) and Examples 10.4, 10.9, 10.10 and 10.11.
5. (i) Use $1/R = 1/R_1 + 1/R_2$ or $R = \dfrac{\text{product}}{\text{sum}}$ [see Examples 10.11 and 10.12] to calculate the equivalent resistance of the 3 Ω and 6 Ω in parallel. Hence calculate the total resistance of the circuit (12 Ω). Use $E = I \times$ (total resistance) [see Section 10.1(a)] to calculate the current (3 A).

   (ii) The current divides in the ratio of the resistances (see Example 10.9), so 2 A passes through the 3 Ω resistor.

   (iii) Use $V = IR$; potential difference = 30 V.

   (iv) Energy dissipated = $I^2 R$ [Section 10.2] = 90 W.

6. See Section 10.1(e), and Examples 10.3 and 10.20.
7. See Section 10.1(f).
8. See Sections 10.4(f) and 11.2. Also Examples 10.14, 10.18 and 10.20.
9. Energy used = (2×3) kWh = 6 kWh. Cost = (6×6)p = 36p.
10. See Section 10.4(b), and Examples 10.16 and 10.19.
11. See Section 10.4(c), and Example 10.15.
12. See Section 10.4, and Examples 10.13, 10.14, 10.19 and 10.20.
13. See Examples 10.14 and 10.15. Also Section 10.4(d).
14. See Section 10.4(e), and Examples 10.20 and 10.21.

10.8 Questions

(Answers and hints on solutions will be found in Section 10.9.)

Question 10.1

(a) A potential divider is constructed as shown in the diagram. What is the reading on the high-resistance voltmeter?

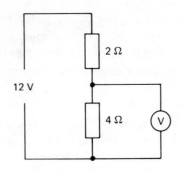

(b) The diagram shows a circuit which could be used to charge a 12 V car battery.

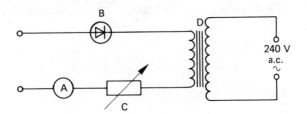

(i) State the name of each of the components labelled A, B, C and D.
(ii) State the function performed by each of the components A, B, C and D. **(10 marks)**

Question 10.2

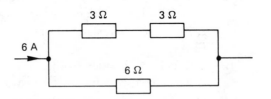

(a) The diagram shows a network of resistors.
   (i) What current flows in the 6 Ω resistor?
   (ii) What charge passes through the 6 Ω resistor in 3 s?
   (iii) What is the p.d. across the 6 Ω resistor?   **(3 marks)**
(b) A 3 kW, 250 V electric fire is connected across a 250 V mains.
   (i) How much current does the fire take?
   (ii) What energy does it consume if left on for 5 hours?
   (iii) How much would it cost to leave it on for 5 hours if the cost of a unit of electrical
        energy (a kWh) is 6p?   **(5 marks)**

Question 10.3

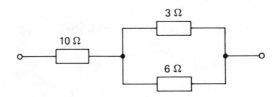

The effective resistance, in Ω, of the above circuit is
A $10\frac{1}{2}$ B 12   C 13   D $14\frac{1}{2}$ E 19   (AEB)

146

Question 10.4

A 1.5 kΩ resistor is connected in series with an LDR and a 10 V d.c. supply. When light shines on the LDR, its resistance is 500 Ω.

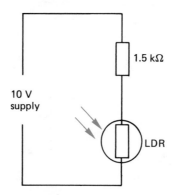

(a) Calculate the current in milliamperes which is passing through the circuit when light is shining on the LDR. **(3 marks)**

(b) What would be the reading on a voltmeter of very high resistance if it were connected across the LDR? **(3 marks)**

(c) What change takes place in the resistance of the LDR when the light is switched off? **(2 marks)**

(d) What would happen to the reading on the voltmeter which is connected across the LDR when the light is switched off? Explain your answer. **(4 marks)**

Question 10.5

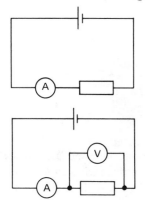

A 1 m length of resistance wire, an ammeter and a 2.0 V cell are connected in series as shown in the figure. (The ammeter and the cell have negligible resistance.)

(a) The ammeter reads 0.50 A. Calculate the resistance of the resistance wire.

(b) The resistance wire is replaced by a 1 m length of wire of the same material but of twice the cross-sectional area. What is
  (i) the resistance of this wire?
  (ii) the new ammeter reading?

(c) A voltmeter is now connected across the resistance wire as shown in the figure. Will the ammeter reading increase, decrease or remain the same? Give a reason for your answer. **(6 marks)**
(L)

Question 10.6

A 2 volt accumulator of negligible internal resistance is connected to an ammeter and three 1 Ω resistors as shown. In which one of the arrangements does the ammeter give the largest reading?

(AEB)

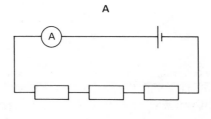

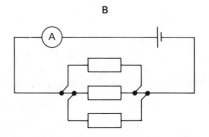

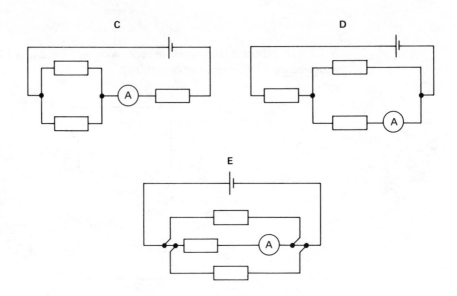

Question 10.7

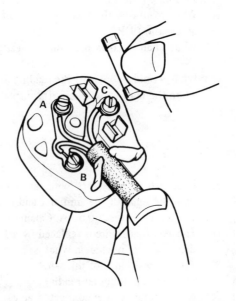

The diagram shows the type of 3-pin electrical plug with the cover removed used in the United Kingdom. The electric cable connected to the plug contains three wires with colour-coded insulation, namely brown, blue and green/yellow stripe. Identify each of the colour-coded wires by stating to which of the terminals **A**, **B** or **C** in the diagram they should be connected. **(2 marks)**

Identify the terminal through which no current passes in normal circumstances. What is the purpose of the wire connecting this terminal to an electrical appliance such as an electric fire? Describe how it works.

What is the purpose of the device, held above C in the diagram, that is about to be inserted into the plug? Describe how it works. **(10 marks)**

(L, part question)

Question 10.8

(a) A sealed box has two terminals on it and nothing else is visible. It may contain any one of the following: a diode, a coil of nichrome wire, a torch bulb.
   (i) Draw the circuit you would use to enable you to obtain readings from which you could draw the current–voltage characteristic for the device that is in the box.

148

(ii) Outline the experiment you would carry out to determine which of the three components is in the box and sketch the graphs you would expect to obtain for each of the three devices. **(13 marks)**

(b) Three fuses are available rated at 2 A, 10 A and 13 A. Which fuse would you choose for an electric fire rated at 3 kW, 250 V? Show how you arrive at your answer. **(4 marks)**

(c) 60 J of heat was dissipated in a resistor when 20 C flowed for 5 s. Calculate
   (i) the potential difference across the resistor,
   (ii) the resistance of the resistor,
   (iii) the average power dissipated in the resistor. **(8 marks)**

Question 10.9

(a) In the circuits shown in Fig. 1 a switch which is open represents the logic state 0 and a switch which is closed represents the logic state 1. The lamp is lit when the output is logic state 1. What type of gate is represented by the circuits in Fig. 1(i) and (ii)? **(4 marks)**

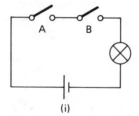
(i)

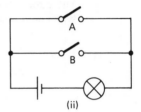
(ii)

Fig. 1

(b) Draw the truth table for the combination of three NOR gates shown in Fig. 2, and state the name of the gate resulting from such a combination. **(4 marks)**

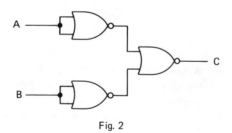

Fig. 2

Question 10.10

The circuit below contains four NAND gates connected together as shown.

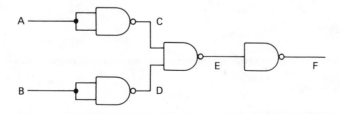

Complete the truth table shown below.

A	B	C	D	E	F
0	0				
0	1				
1	0				
1	1				

What is the resulting gate called?
Show how three **NOR** gates may be connected so that they form an **AND** gate and write the truth table for the arrangement.

Question 10.11

The circuit diagram shows a circuit for automatically switching on lights when it gets dark. When light is shining on the LDR, its resistance is 600 Ω.

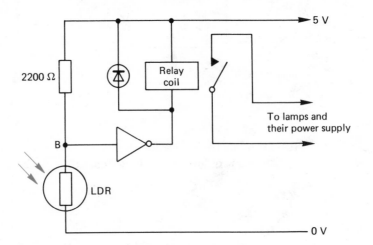

(a) What is the voltage at B when the resistance of the LDR is 550 Ω?   (3 marks)
(b) When it is dark, the resistance of the LDR is 220 kΩ. What is the voltage at B when it is dark?   (3 marks)
(c) Explain what happens in this circuit when the light shining on the LDR goes out.   (4 marks)
(d) What is the purpose of the diode in the circuit?   (2 marks)

Question 10.12

Which of the circuits below could be used to arrange for the lamp to light about 5 s after the switch S is closed?

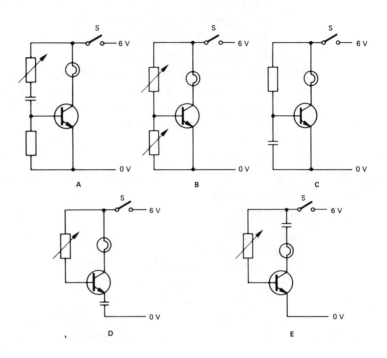

Question 10.13

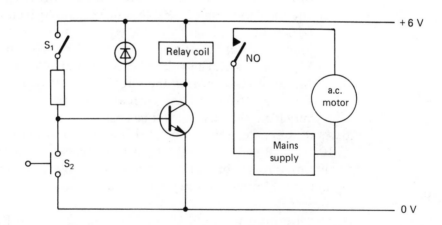

The circuit shown in the diagram is designed to switch the motor on and off.

(i) With the switches in the position shown is
   (a) the motor on or off?   **(2 marks)**
   (b) a current flowing in the collector of the transistor?   **(2 marks)**
(ii) Switch S_1 is closed and switch S_2 left open. Is the motor on or off? Explain your answer.   **(4 marks)**
(iii) S_2 is now closed (S_1 being left closed). What changes take place
   (a) in the current in the collector?
   (b) in the operation of the motor?   **(3 marks)**
(iv) The switch S_2 is replaced by an LDR. Will the motor be on or off
   (a) in bright daylight?
   (b) in darkness?
      Explain your answers.   **(4 marks)**

10.9 Answers and Hints on Solutions to Questions

1. (a) Current in resistors = 2 A. Use $V = IR$ [see Section 10.1] = 2 A $\times$ 4 Ω = 8 V.

   (b) (i) A is an ammeter, B a diode, C a variable resistor and D a step-down transformer. (ii) A records the current; B will only pass current in one direction; C enables the current to be adjusted; D reduces the potential difference to about 15 V.

2. (a) The arrangement is equivalent to two 6 Ω resistors in parallel. (i) The current divides equally, 3 A flowing in each branch. (ii) 3 C/s; hence, in 3 s charge is 9 C. (iii) Use $V = IR$; potential difference = 18 V.

   (b) (i) Use power (watts) = potential difference (V) $\times$ current (I) [Section 10.2]; current 12 A.

   (ii) 3000 J/s; in 5 hours, 54 MJ. (iii) Cost = kWh $\times$ 6p = (3 $\times$ 5 $\times$ 6)p = 90 p.

3. Use $1/R = 1/R_1 = 1/R_2$ to get equivalent resistance of 3 Ω and 6 Ω in parallel. Total resistance = (10 + 2) Ω = 12 Ω.

   Answer **B**

4. (a) $I = \dfrac{V}{R}$ [see Section 10.1]

   Total resistance of circuit = 1500 Ω + 500 Ω = 2000 Ω

   Hence, $I = \dfrac{10 \text{ V}}{2000 \ \Omega}$ = 0.005 A = 5 mA

   (b) p.d. across LDR = IR = 0.005 A $\times$ 500 Ω = 2.5 V

   (c) The resistance of the LDR goes up.

   (d) The voltmeter reading goes up, because the current in the circuit goes down and the voltage drop across the 1.5 kΩ resistor goes down. Hence, the voltage drop across the LDR goes up, because the sum of the voltage drops across the 1.5 kΩ resistor and the LDR is 10 V.

5. (a) Use the definition of resistance in Section 10.1(a). Resistance = 4 Ω.

   (b) (i) The new wire has twice the cross-sectional area and this is equivalent to joining two of the original wires in parallel. New resistance = 2 Ω. (ii) 1 A.

   (c) Slight increase. The voltmeter is in parallel with the resistor, and, hence, the total resistance of the circuit is decreased (see Section 10.1(d) and Example 10.5) and the current increases. If the voltmeter resistance is very high, the increase will be negligible.

6. As resistors are added in parallel, the total resistance goes down (see Section 10.1(d)). **B** and **E** have the lowest resistance. The ammeter in **E** registers the current through one of the resistances, while the ammeter in **B** registers the sum of the currents through the three resistors.

   Answer **B**

7. The live brown wire to C, the blue neutral wire to B, and the earth yellow/green striped wire to A. No current passes in A in normal circumstances (see Section 10.3).

8. (a) The experiment is described in Example 10.11. Alternatively, the circuit in Example 10.5 could be used. The graph for a torch bulb is drawn in Example 10.5. The graph for the coil will be a straight line through the origin (it obeys Ohm's law); the diode will be discovered by turning the box round and connecting the opposite ends to the battery. In one direction there will be a very high resistance and very little current will pass. For negative (reversed) voltages the current will be very small and the graph is very nearly horizontal just below the voltage axis.

(b) Current is 12 A and, hence, the 13 A fuse will be needed.

(c) (i) Voltage is $\dfrac{J}{C}$ (see Section 10.1); hence, potential difference is 3 V.

(ii) Current is 4 A; hence, resistance is 0.75 Ω. (iii) Power is (60 J)/(5 s) = 12 W.

9. (a) (i) is an AND and (ii) is an OR gate.

(b) You must draw the truth table. The combination is an AND gate.

10.

A	B	C	D	E	F
0	0	1	1	0	1
0	1	1	0	1	0
1	0	0	1	1	0
1	1	0	0	1	0

The above gate is a NOR gate.

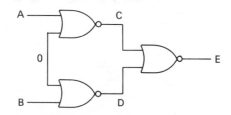

A	B	C	D	E
0	0	1	1	0
0	1	1	0	0
1	0	0	1	0
1	1	0	0	1

11. (a) 1 V.

(b) 4.95 V.

(c) B's voltage goes from 1 V to 4.95 V and the output from the NOT gate goes from high to low. A current flows in the relay coil which operates the switch and lights the lamps.

(d) It protects the NOT gate when the current in the relay drops to zero and there is a large induced e.m.f. (see Section 11.4). The resulting induced current could damage the NOT gate.

12. Answer C

The lamp will light when a current passes through the base–emitter circuit of the transistor and switches the transistor on. In C this will happen after the capacitor has charged up.

13. (i) (a) off; (b) no.

(ii) On. The current in the base–emitter circuit of the transistor switches the transistor on. The current in the collector circuit operates the relay and the motor switches on.

(iii) When S_2 is closed, the transistor switches off. (a) There is no current in the collector circuit and the relay switch is open. (b) The motor is off.

(iv) (a) In daylight the resistance of the LDR will be low and the transistor will be switched off. In darkness the resistance of the LDR will be high, the transistor will be switched on and the motor will be on.

11 Magnetism, Electromagnetism, Motors, Dynamos and Transformers

11.1   Magnetism

Certain substances, including iron, steel, cobalt and nickel, are magnetic.

The region of space around a magnet where its magnetic influence may be detected is known as a magnetic field, and contains something we call magnetic flux. The flux patterns may be shown by sprinkling iron filings around a magnet and gently tapping the surface, or by using plotting compasses. The flux patterns resulting from a bar magnet and a U-shaped magnet are shown in Fig. 11.1.

Magnets attract unmagnetised pieces of iron and steel. To discover whether a bar of iron or steel is magnetised, bring each end up to the N-pole of a suspended magnet. If one end repels the N-pole, the bar is magnetised (like magnetic poles repel each other). If both ends attract the N-pole, the bar is not magnetised.

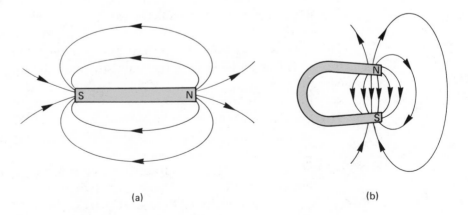

(a)   (b)

Figure 11.1   The lines of flux (a) due to a bar magnet and (b) due to a U-shaped magnet.

154

A bar of magnetic material may be magnetised by putting it in a solenoid carrying a current. It may be demagnetised by withdrawing it slowly from a solenoid in which an alternating current is passing. Recording heads in a tape recorder use solenoids to magnetise the tape.

Materials such as steel and alcomax (a steel-like alloy) are difficult to magnetise and are said to be *hard*. Iron and mumetal (a nickel alloy) are relatively easy to magnetise, but they do not retain their magnetism, and are said to be *soft*.

11.2 Electromagnetism

When a conductor carries a current it produces a magnetic field. The flux patterns resulting from a current passing through a straight wire and a solenoid are shown in Fig. 11.2.

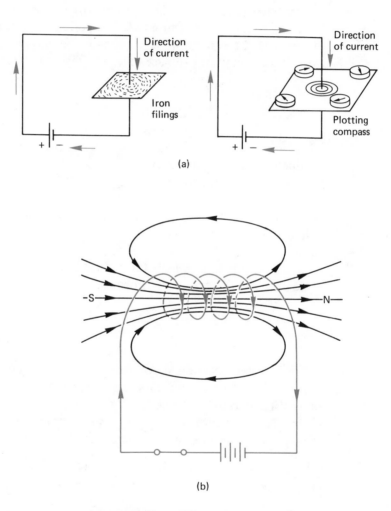

Figure 11.2 The flux patterns caused by (a) a current passing through a straight wire and (b) a current passing through a solenoid.

The direction of the field may be found by use of the 'screw rule', which states that if a right-handed screw is turned so that it moves forward in the same direction as the current, its direction of rotation gives the direction of the magnetic field.

The effect is used in electromagnets, which are usually solenoids wound round iron or mumetal. Iron and mumetal are soft magnetic materials that quickly lose their magnetism when the current in the solenoid is switched off.

Electromagnetic relays use the magnetic effect of a current passing through a solenoid to operate a switch. The 'pull on' current is the value of the current in the solenoid needed to close the switch, and the 'switch off' current is the value of the current when the switch opens.

11.3 The Electric Motor and the Dynamo

See also Examples 11.3 and 11.7. Electric motors and dynamos basically consist of a coil which can rotate in a magnetic field. In the motor a current passes through the coil, and the resulting forces on the coil cause it to rotate. A split-ring commutator reverses the direction of the current every half-revolution, thus ensuring continuous rotation. The turning effect may be increased by increasing the current, increasing the number of turns on the coil or increasing the strength of the magnetic field. In the dynamo the coil is rotated, and as the wires cut the lines of flux, an e.m.f. (an e.m.f. is a voltage: see Section 10.1) is induced across the ends of the coil. In an a.c. dynamo the ends of the coil are connected to slip rings. If a commutator replaces the slip rings, the output is d.c.

11.4 Faraday's Law of Electromagnetic Induction

Whenever an e.m.f. is induced in a conductor due to the relative motion of the conductor and a magnetic field, the size of the induced e.m.f. is proportional to the speed of the relative motion.

Alternatively, the law may be stated:
Whenever there is a change of magnetic flux linked with a circuit an e.m.f. is induced. The e.m.f. is proportional to the rate of change of flux linked with the circuit.

If the circuit is complete, the induced e.m.f. produces a current. The effect is made use of in dynamos, transformers and playback heads of tape recorders.

11.5 Transformers

A primary and secondary coil are wound on a continuous core made of a soft magnetic alloy (cores of modern alloys have largely replaced iron cores). The alternating current passing through the primary causes a changing flux in the secondary and hence an induced e.m.f. across the secondary coil. Most modern transformers are made so that there is negligible magnetic flux loss and

$$\frac{\text{voltage across secondary}}{\text{voltage across primary}} = \frac{\text{number of turns on secondary}}{\text{number of turns on primary}}$$

$$\text{efficiency} = \frac{\text{power out}}{\text{power in}} = \frac{(VI)_{\text{secondary}}}{(VI)_{\text{primary}}}$$

Transformers have high efficiencies, and in many cases a good approximation is

$$(VI)_{\text{secondary}} = (VI)_{\text{primary}}$$

As in all appliances, the switch and fuse must be in the live wire.

Electricity is transmitted across the British countryside on the National Grid. High voltages are used, as low-voltage power lines are wasteful and inefficient, because of the large currents needed. The energy lost as heat in the wires increases rapidly as the current increases (energy lost as heat = $I^2 Rt$: see Section 10.2).

11.6 Left-hand Rule (for the motor effect)

Hold the first finger, second finger and thumb of the left-hand mutually at right angles, so that the *First* finger points in the direction of the magnetic *Field*, the se*C*ond finger in the direction of the *C*urrent, then the *TH*u*M*b points in the direction of the *TH*rust or *M*otion.

11.7 Right-hand Rule (for the dynamo effect)

Hold the first finger, second finger and thumb of the right-hand mutually at right angles, so that the *First* finger points in the direction of the *F*ield, the thu*M*b in the direction of the *M*otion, then the se*C*ond finger points in the direction of the induced *C*urrent.

11.8 Worked Examples

Example 11.1

(a) (i) Explain why iron rather than hard steel is used in electromagnets. (ii) State two ways of increasing the strength of an electromagnet. **(4 marks)**

(b) Two pieces of steel, A and B, are lying on a bench. The N-pole of a magnet when brought up to one end of A attracts it, and when brought up to one end of B repels it. What can you deduce about the state of magnetism of A and B?

Solution 11.1

(a) (i) Iron is more easily magnetised than hard steel and loses its magnetism more easily. If the magnetising current were switched off, the iron would lose its magnetism but the steel would retain some magnetism. (ii) Increase the number of turns. Increase the current.

(b) B is magnetised. A may be magnetised. The end of B to which the magnet was brought up is a N-pole.

[Repulsion is a test for magnetism. If the N-pole of the magnet were to be brought up to the S-pole of B, then attraction would occur. If B were not magnetised, it would still be attracted by the magnet, because the N-pole of the magnet would induce an S-pole on B. The N-pole and the S-pole would attract each other. Attraction is not a test for magnetism, so A may or may not be magnetised.]

Example 11.2

When the key in the circuit shown is closed, the bare wire XY will
A rise up off the bare wires
B press down more strongly
C move to the left
D move to the right
E remain stationary

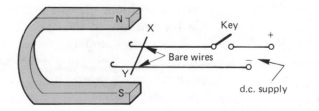

(AEB)

Solution 11.2

[Use the left-hand rule. *F*irst finger for *F*ield, se*C*ond finger for *C*urrent (+ to −) and thu*M*b for *M*otion. The wire will move to the right.]

Answer **D**

Example 11.3

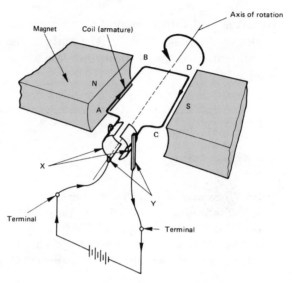

The coil is wound on a soft iron core (not shown in the diagram).

The diagram shows a d.c. electric motor.
(a) What is the part X called?   **(1 mark)**
(b) What is the part Y called?   **(1 mark)**
(c) Explain why the coil of the motor rotates continuously when the motor is connected to a d.c. supply.   **(7 marks)**
(d) Suggest three ways in which the motor could be made to go faster.   **(3 marks)**
(e) What would happen to the rotation if when the direction of the current in the coil were reversed at the same time the direction of the field were reversed?   **(2 marks)**
(f) If the motor is 3 W, 6 V and its efficiency is 20%, what is
  (i) the current it takes?
  (ii) the energy it can supply in 10 s?   **(6 marks)**

Solution 11.3

(a) Split-ring commutator.
(b) Brushes.
(c) The wires AB and CD are current-carrying conductors and they are in a magnetic field which is perpendicular to the wire. Each side of the coil has a force acting on it. The force on AB is downwards and the force on CD is upwards. This couple [two equal and opposite parallel forces] would cause the coil to rotate through 90° from the position shown. As the coil passes through the vertical position, the two sections of the split-ring commutator are connected, via the brushes, to the opposite terminals of the battery. The current in the coil is reversed and the forces acting on the sides of the coil are reversed in direction. The forces rotate the coil through 180° until it is next in the vertical position, when the current and the forces are again reversed. The coil thus continues to rotate.

(d) Increase the current. Increase the strength of the magnetic field. Increase the number of turns on the armature coil.

(e) If both the current and the field were reversed at the same time, the forces would continue to act in the same direction and the coil would not rotate continuously.

(f) (i) Power (watts) = potential difference (volts) × current (amperes) [Section 10.2]

$$3 \text{ W} = 6 \text{ V} \times \text{current}$$
$$\Rightarrow \quad \text{current} = \frac{3 \text{ W}}{6 \text{ V}} = 0.5 \text{ A}$$

(ii) [The 3 W is the power taken in. We need to know the power got out, so we use the equation in Section 4.4.]

$$\text{Efficiency} = \frac{\text{power out}}{\text{power in}}$$

[Remember to put the efficiency as a fraction or decimal, NOT as a percentage.]

$$0.2 = \frac{\text{power out}}{3 \text{ W}}$$

$$\text{Power out} = 0.2 \times 3 \text{ W} = 0.6 \text{ W} = 0.6 \frac{\text{J}}{\text{s}}$$

So, in 10 s the energy supplied $= 0.6 \frac{\text{J}}{\text{s}} \times 10 \text{ s} = 6 \text{ J}$.

Example 11.4

The diagram represents a transformer with a primary coil of 400 turns and a secondary coil of 200 turns.

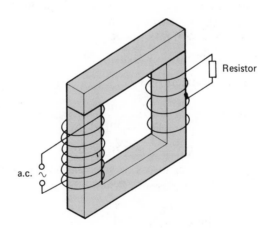

(a) If the primary coil is connected to the 240-V a.c. mains what will be the secondary voltage? **(1 mark)**

(b) Explain carefully how the transformer works. **(4 marks)**

(c) Calculate the efficiency of the transformer if the primary current is 3 A and the secondary current 5 A. **(4 marks)**

(d) Give reasons why you would expect this efficiency to be less than 100%. **(3 marks)**

(e)

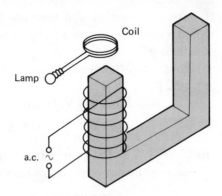

The secondary coil is removed and a small coil connected to a low voltage lamp is placed as shown. Explain the following observations:

  (i) the lamp lights, **(2 marks)**

 (ii) if the coil is moved upwards, the lamp gets dimmer, **(2 marks)**

(iii) If an iron rod is now placed through the coil, the lamp brightens again, **(2 marks)**

(iv) the lamp will not light if a d.c. supply is used instead of an a.c. one. **(2 marks)**

                                                             (L)

Solution 11.4

(a) 120 V   [This assumes that there are no flux losses.]

(b) When an a.c. is connected to the primary coil there is a constantly changing magnetic flux in the iron core. This continuously changing flux passes through the secondary coil, and hence there is an induced e.m.f. in the secondary coil. Since the same e.m.f. is induced in each turn of the secondary coil and all the turns are in series, halving the number of turns of the secondary coil halves the induced e.m.f.

(c) Power input = $(VI)_{primary}$ = 240 V × 3 A = 720 W   [see Section 10.2]

Power output = $(VI)_{secondary}$ = 120 V × 5 A = 600 W

Efficiency = $\dfrac{\text{power output}}{\text{power input}}$ [see Section 11.5] = $\dfrac{600 \text{ W}}{720 \text{ W}}$ = 0.83 = 83%

(d) The efficiency is less than 100% because of

  (i) Copper losses, that is the heat produced in the wires.

 (ii) Eddy currents flowing in the iron core produce heat.

(iii) Magnetic leakage. Not all the flux in the iron on the primary side reaches the secondary side.

(iv) The work done in continually magnetising and demagnetising the iron core (this is called the hysteresis loss).

(e) (i) Some of the magnetic flux around the primary passes through the coil. This changing flux means that an e.m.f. is induced in the coil. This e.m.f. produces a current in the coil which lights the lamp.

 (ii) When the coil is moved further away from the primary, the flux passing through it decreases. The rate of change of flux is less and the magnitude of the induced e.m.f. is reduced.

(iii) The iron rod will become magnetised and will increase the flux passing through the coil. The rate of change of flux is increased and hence the e.m.f. is increased.

(iv) When a d.c. supply is used the flux through the coil is constant. If there is no change of flux there is no induced e.m.f.

Example 11.5

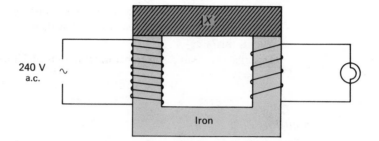

240 V a.c.

Iron

The circuit shown in the diagram was set up in order to demonstrate a step-down transformer. The lamp glowed dimly. The lamp would glow more brightly if

A the number of turns on the primary coil were reduced
B the iron were replaced by copper
C the shaded section of iron, X, were removed
D the number of turns on the secondary were reduced

Solution 11.5

[The voltage across the secondary coil depends on the turns ratio (Section 11.5). Reducing the number of turns on the primary coil increases the turns ratio and, hence, increases the voltage across the secondary coil. Reducing the number of turns on the secondary coil reduces the turns ratio and, hence, reduces the voltage across the secondary coil. Copper is non-magnetic and so is not suitable for the core. The flux through the secondary coil will be reduced if the magnetic circuit is not complete, so removing X will decrease the voltage across the secondary coil.]

Answer **A**

Example 11.6

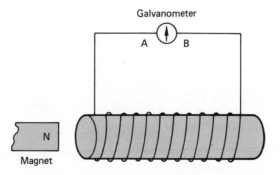

Galvanometer

A   B

N

Magnet

(a) The diagram shows a solenoid connected to a galvanometer. Explain why

(i) if the magnet is held stationary at the end of the coil, there is no deflection of the galvanometer pointer,

(ii) if the magnet is moved towards the solenoid there is a deflection of the galvanometer pointer,

(iii) the faster the magnet moves towards the solenoid the greater is the deflection of the galvanometer pointer. **(6 marks)**

(b) A transformer has 400 turns in the primary winding and 10 turns in the secondary winding. The primary e.m.f. is 250 V and the primary current is 2.0 A. Calculate

  (i) the secondary voltage, and

  (ii) the secondary current, assuming 100% efficiency. **(6 marks)**

Transformers are usually designed so that their efficiency is as close to 100% as possible. Why is this?

Describe *two* features in transformer design which help to achieve high efficiency.

(4 marks)

Solution 11.6

(a) (i) If the magnet is stationary the flux through the coil is not changing. An e.m.f. will only be induced in the coil if the flux through it is changing.

(ii) When the magnet is moved towards the solenoid the flux through the coil is changing and hence there is an induced e.m.f.

(iii) The magnitude of the induced e.m.f. depends on the rate of change of flux. If the magnet is moved faster the rate of change of flux is greater and hence the induced e.m.f. increases and the resulting current increases.

(b) (i) $\dfrac{\text{Voltage across secondary}}{\text{Voltage across primary}} = \dfrac{\text{number of turns on secondary}}{\text{number of turns on primary}}$

[see Section 11.5]

$$\frac{\text{Voltage across secondary}}{250 \text{ V}} = \frac{10 \text{ turns}}{400 \text{ turns}}$$

Voltage across secondary = 6.25 V

(ii) $(VI)_{\text{primary}} = (VI)_{\text{secondary}}$ [see Section 11.5]

$250 \text{ V} \times 2.0 \text{ A} = 6.25 \text{ V} \times I_{\text{secondary}}$
$I_{\text{secondary}} = 80 \text{ A}$

The greater the losses the greater is the cost of running the transformer and the hotter the transformer becomes when being used. Large transformers have to be cooled and this can be expensive. The core is laminated, that is it is made up of strips insulated from each other. This reduces the eddy currents. The coils are wound using low resistance material so that the heating effect in the coils is reduced to a minimum.

Example 11.7

The diagram shows a simple form of a.c. dynamo.

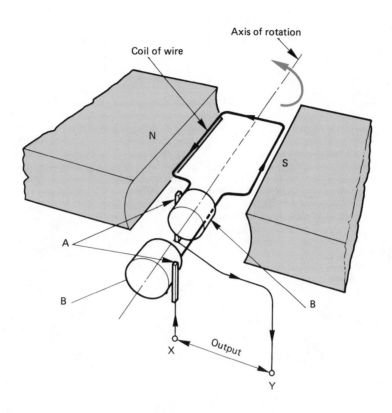

(a) (i) What are the names of the parts labelled A and B?
   (ii) What would be the effect of doubling the number of turns on the coil if the speed of rotation remained unchanged?
   (iii) Which of the output terminals is positive if the coil is rotating in the direction shown in the diagram (anticlockwise)? **(8 marks)**
(b) What is the position of the rotating coil when the p.d. across its ends is zero? Explain why the p.d. is zero when the coil is in this position. **(3 marks)**
(c) Sketch a graph showing how the p.d. across the ends of the rotating coil varies with time for an a.c. dynamo. On the same sheet of paper and vertically below the first graph and using the same time scale, sketch graphs to show the effect of (i) doubling the speed of rotation and at the same time keeping the field and the number of turns constant; (ii) doubling the number of turns on the coil and at the same time doubling the speed of rotation of the coil, keeping the field constant. **(9 marks)**

Solution 11.7

(a) (i) The brushes are labelled A and the slip rings labelled B.
   (ii) The induced e.m.f. would double.
   (iii) Y is the positive terminal.
(b) The p.d. across the ends of the coil is zero when the coil is in a vertical plane. In this position the sides of the coil are moving parallel to the lines of flux and the flux through the coil is not changing; hence, there is no induced e.m.f.

(c)

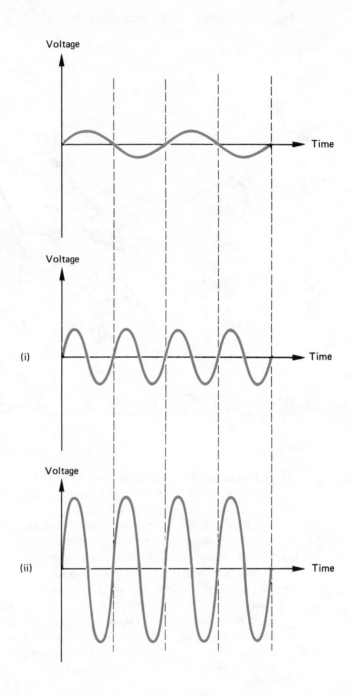

[In (ii) the voltage goes up by a factor of 4 because doubling the number of turns doubles the voltage, and doubling the speed of rotation also doubles the voltage. Doubling the speed of rotation also *doubles the frequency*.]

Example 11.8

(a)

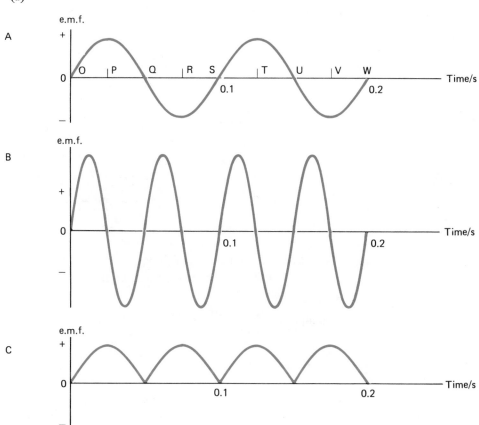

Graph A above shows how the e.m.f. produced by a simple dynamo varies with time. Graphs B and C show how the e.m.f. produced by the same dynamo varies with time after certain alterations and modifications have been made.

(i) How many revolutions has the coil of the dynamo made in the time interval OT on graph A? **(1 mark)**

(ii) What is the frequency of the alternating e.m.f. as shown by graph A? **(2 marks)**

(iii) Which letters on graph A correspond to the plane of coil of the dynamo being parallel to the magnetic field? **(2 marks)**

(iv) Explain why the e.m.f. at Q is zero. **(3 marks)**

(v) What alteration has been made for the dynamo to produce the e.m.f. represented by graph B? **(2 marks)**

(vi) What modification has been made to the dynamo for it to produce the e.m.f. represented by graph C? Illustrate your answer with sketches showing the original and the modified arrangements. **(4 marks)**

(b) A dynamo is driven by a 5-kg mass which falls at a steady speed of 0.8 m/s. The current produced is supplied to a 12-W lamp which glows with normal brightness. Calculate the efficiency of this arrangement. **(6 marks)**

(L)

Solution 11.8

(a) (i) 1.25.

(ii) 1 cycle takes 0.1 s. Frequency = 10 Hz.

(iii) P; R; T; V. The e.m.f. is a maximum because the wire on the sides of the coil is cutting the field at the greatest rate.

(iv) At Q the plane of the coil is perpendicular to the magnetic field. The sides of the coil are moving parallel to the field, so they are not cutting the lines of flux and the rate of change of flux is zero.

(v) The speed of rotation has been doubled.

(vi) Slip rings have been replaced by split rings (see diagram).

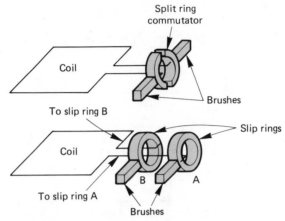

(b) Work done in 1 s by mass falling = force × distance travelled in 1 s = 50 N × 0.8 m/s = 40 W [The force on the 5 kg mass is 50 N.]

Work got out = 12 W

Efficiency = $\dfrac{\text{work out}}{\text{work in}}$ = $\dfrac{12}{40}$ = 0.3 or 30%

Example 11.9

(i) The diagram below shows a moving coil loudspeaker. Make a sketch of the magnet and show on your sketch the nature and the position of the magnetic poles.

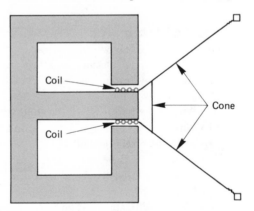

(ii) Explain what would be heard if a 1 V, 50 Hz supply were connected across the terminals of the loudspeaker. What difference would be heard if the supply were changed to a 2 V, 100 Hz one? **(8 marks)**

Solution 11.9

(i)

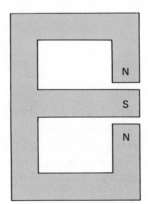

166

(ii) The wire of the coil is in a magnetic field. When a current flows in the wire of the coil a force acts on the wire. This force is perpendicular to both the field and the current. When an alternating current is flowing in the coil, the coil will move to the left and then to the right each time the current changes direction. The cone will move in and out fifty times every second and a note of 50 Hz will be heard.

If the supply were changed to 2 V, 100 Hz, the amplitude of the vibration would be greater, so a louder note would be heard. The cone would complete 100 cycles every second and a note of 100 Hz would be heard.

Example 11.10

(a) Explain why on the National Grid system
 (i) very high voltages are used,
 (ii) alternating current is used. **(5 marks)**
(b) A town receives electricity via the National Grid system at 100 kV at a rate of 40 MW. The cable connecting the town to the power station has a total resistance of 4 Ω. What is
 (i) the current passing through the cable?
 (ii) the power loss as a result of heating in the cable? **(5 marks)**

Solution 11.10

(a) (i) A certain amount of energy may be transmitted by using a high voltage and a low current, or a low voltage and a high current. The former is chosen because the heating effect depends on the square of the current and the loss as heat is reduced markedly by using a high voltage. Low current also means that the wires do not have to be so thick and the cost of supporting the wires is reduced.
 (ii) Using alternating current means that transformers can be used to step up the voltage before transmission and also they reduce it again at the end of the power line.
(b) (i) Power = (p.d.) × (current) [see Section 10.2]
 40×10^6 W = 100×10^3 V × current
 current = 400 A
 (ii) Power loss = $I^2 R$ watt [see Section 10.2]
 = $(400^2 \times 4)$ W = 6.4×10^5 W

Example 11.11

Calculators may be powered by the mains using an adaptor or by dry batteries.
 (i) Why is an adaptor necessary if the mains is used? **(2 marks)**
 (ii) Outline the principle by which the adaptor works (no details of the electrical circuit or a diagram are necessary). **(4 marks)**

Solution 11.11

(i) Calculators run off about 6 V and the mains is 240 V. The adaptor reduces the mains voltage before supplying the calculator.

(ii) The adaptor has a transformer which consists of a primary coil and a secondary coil wound on a magnetic alloy. The 240 V is connected across the primary coil. This alternating voltage causes a changing flux in the soft magnetic alloy, and this changing flux passes through the secondary coil. The changing flux in the secondary coil results in an e.m.f. being induced in it. The secondary coil has only one-fortieth of the turns the primary coil has on it and hence the voltage is reduced by one-fortieth (if 6 V is required). A rectifier converts the a.c. voltage to a d.c. voltage.

11.9　Have You Mastered the Basics?

1. Can you describe how to identify magnetic poles?
2. Can you describe an experiment to obtain the flux patterns around a bar magnet, a straight wire carrying a current and a solenoid carrying a current?
3. Can you describe how to use a solenoid for magnetising and demagnetising a material?
4. Do you understand Faraday's law of electromagnetic induction?
5. Can you draw a diagram of a motor and a dynamo, and describe how they work?
6. Can you draw a diagram of a transformer and describe how it works?
   Do you know how to calculate the efficiency of a transformer?
7. A transformer reduces a mains voltage of 240 V to 6 V. If the transformer is 100% efficient what is
   (i) the turns ratio?
   (ii) the current drawn from the mains if the current in the secondary is 1 A?

11.10　Answers and Hints on Solutions to 'Have You Mastered the Basics?'

1. See Section 11.1.
2. See Sections 11.1 and 11.2.
3. See Section 11.1.
4. See Section 11.4 and Example 11.6.
5. See Section 11.3 and Examples 11.3, 11.7 and 11.8.
6. See Section 11.5 and Examples 11.4 and 11.5.
7. $\dfrac{\text{Number of turns on secondary}}{\text{Number of turns on primary}} = \dfrac{\text{voltage across secondary}}{\text{voltage across primary}}$ [see Section 11.5]

$$= \frac{6}{240} = \frac{1}{40}$$

Turns ratio = 1 : 40
The transformer is 100% efficient; therefore
power in = power out
$(VI)_{\text{primary}} = (VI)_{\text{secondary}}$　　　　[see Section 11.5]
$240 \text{ V} \times I_{\text{primary}} = 6 \text{ V} \times 1 \text{ A}$
$I_{\text{primary}} = 0.025 \text{ A}$

11.11 Questions

(Answers and hints on solutions will be found in Section 11.12.)

Question 11.1

This question is about magnetic fields and an electromagnetic switch.

(a) (i) Describe how you would show experimentally the shape and direction of the magnetic field lines in a horizontal plane around a vertical wire connected to a d.c. source.
   (ii) Draw a diagram showing clearly the direction of the current and the direction of the magnetic field lines.
   (iii) If a.c. were used in place of d.c., what effect would this have on your experiment? Give a reason. **(8 marks)**

(b)

The diagram shows a small plotting compass placed between two strong magnets. The tip of the arrow represents the N pole of the compass.
(i) What is the polarity of the end C of the right hand magnet?
(ii) Draw a diagram of the magnets only as seen from above and sketch the magnetic field lines in the region between B and C. **(3 marks)**

(c)

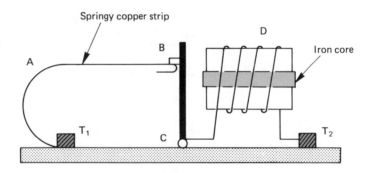

The diagram shows a model circuit breaker designed to switch off the current in a circuit when it becomes excessive. The current enters the circuit breaker at T_1, passes along the copper strip A, the iron armature BC, the coil D, and leaves at terminal T_2. The iron armature BC is pivoted at C.
(i) Describe how the circuit breaker works.
(ii) State the effect on the operation of the circuit breaker of each of the following changes. Give a reason in each case.
   (1) The removal of the iron core from the coil.
   (2) The use of a.c. instead of d.c. **(6 marks)**
   (L)

Question 11.2

(a)

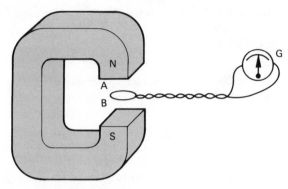

Figure 1

Figure 1 represents a flat circular coil **AB** placed between the poles of a strong magnet and connected to a sensitive galvanometer, **G**. Explain carefully why
  (i) in the position shown, with the side **A** of the coil uppermost, there will be no deflection of the galvanometer, **(2 marks)**
  (ii) when the coil is rotated quickly through 180° so that side **B** of the coil is uppermost there will be a deflection, **(2 marks)**
  (iii) when the coil is pulled quickly away from the magnet there will be a deflection. **(2 marks)**

(b)

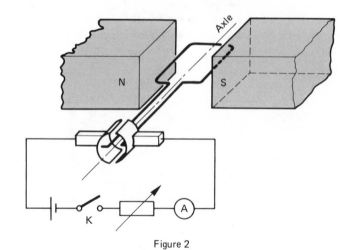

Figure 2

Figure 2 represents a coil, the ends of which are connected to a split ring commutator.
  The coil is placed between the poles of a strong magnet and connected by brushes to the circuit shown. Explain why, when the key **K** is closed,
  (i) the coil starts to rotate, and give the direction of rotation, **(3 marks)**
  (ii) the coil continues to rotate. **(3 marks)**

(c)

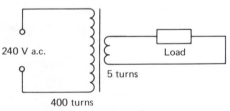

Figure 3

Figure 3 represents a transformer with 400 turns in the primary winding and 5 turns in the secondary winding. The primary e.m.f. is 240 V and the primary current is 2 A.
  (i) What is the secondary voltage? **(2 marks)**
  (ii) Assuming no power loss, what is the secondary current? **(2 marks)**

(iii) Energy is wasted in the transformer. Give *two* reasons why your calculated value in (ii) will be larger than the true value of the current. **(2 marks)**

(iv) Suggest a purpose for which this transformer may be used. **(2 marks)**

(L)

Question 11.3

(a)

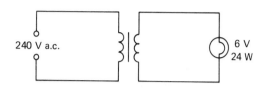

It is required to run a 6-V, 24-W lamp from a 240-V a.c. mains using a transformer as shown above.

(i) Calculate the current that would be taken by the lamp when operating normally.
(2 marks)

(ii) Calculate the turns ratio of the transformer you would use. **(2 marks)**

(iii) Calculate the current taken by the primary coil of the transformer, assuming it to be 100% efficient. **(2 marks)**

(iv) Why, in practice, is the efficiency of the transformer less than 100%? **(3 marks)**

(b) Alternatively the 6-V, 24-W lamp can be operated normally from a 240-V d.c. supply using a suitable fixed resistor, R, as in the diagram.

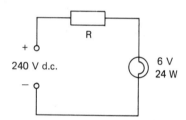

(i) What is the resistance of the lamp? **(2 marks)**

(ii) What is the p.d. across the resistor? **(2 marks)**

(iii) What is the resistance of the resistor? **(2 marks)**

(iv) How much energy is dissipated in the resistor in 1 s? **(2 marks)**

(c) Why may the method used to light the lamp described in (a) be preferable to that described in (b)? **(3 marks)**

(L)

Question 11.4

The diagram shows a step-down transformer with three tappings on the secondary.

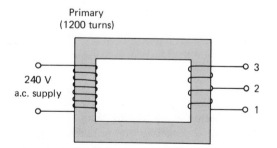

(a) Describe how a transformer works. **(7 marks)**

(b) Why is the efficiency less than 100%? **(5 marks)**

(c) When a 2 Ω resistor is connected between 1 and 2 the current flowing is 0.4 A. When

the same resistor is connected between 2 and 3 the current is 0.6 A. If the transformer is 100% efficient calculate
 (i) the potential difference across terminals 1 and 2,
 (ii) the potential difference across terminals 1 and 3,
(iii) the number of turns on the secondary coil between terminals 1 and 2,
(iv) the number of turns on the secondary coil between terminals 1 and 3.   **(8 marks)**

Question 11.5

A magnet is moved steadily towards a solenoid as shown in the diagram. The ends of the solenoid are connected to a low-resistance centre-zero galvanometer. As the N-pole of the magnet is moved towards the solenoid, the galvanometer deflects to the right.

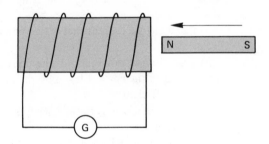

(a) What will the deflection on the galvanometer be when
 (i) the magnet is moved more quickly towards the solenoid?   **(1 mark)**
 (ii) the magnet is turned round so that the S-pole faces the solenoid and the magnet is moved away from the solenoid?   **(2 marks)**
 (iii) the magnet is held stationary with half the magnet inside the solenoid?   **(1 mark)**
(b) How will the deflection on the galvanometer alter when the N-pole is moving towards the solenoid if
 (i) a high resistance is placed in series with the galvanometer?   **(1 mark)**
 (ii) a high resistance is placed in parallel with the galvanometer? Explain your answer.   **(2 marks)**
(c) Outline the physical principles which enable a dynamo to generate a current. (No diagram is necessary.)   **(5 marks)**

Question 11.6

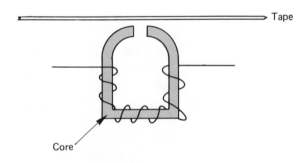

The diagram above shows a recording head to be found in any cassette tape recorder. Speech or music is converted into an alternating current which passes through a wire coiled around the core which makes up the recording head. The tape becomes magnetised.
(a) The tape is usually made of plastic and coated with a thin layer of iron oxide. Why is iron oxide used? On which side of the tape should the coating be?
(b) From what sort of material should the core be made? Why?

(c) Explain why a current through the wire causes the tape to become magnetised.

(d) Describe *three* different ways in which you might change the design to get a larger signal recorded on the tape for the *same* current through the wire.

(e) On playback, another head just like the one above but working in the opposite sense is used to generate a current from the signal recorded on the tape. Someone suggests that this could be explained by saying that the head was acting as a sort of dynamo. Using what you know about how a dynamo works, describe the basic physics principles involved in the playback generation of a current.

(f) If the tape moves at 4.7 cm/s, what length of tape is required to last for 45 min?

(g) A manufacturer wants to make a tape which lasts for 60 min but he cannot simply put more tape in the cassette, because it is already full. Suggest a possible solution. What problem might your solution bring?

11.12 Answers and Hints on Solutions to Questions

1. (a) (i) Pass a vertical wire through a horizontal card with plotting compasses on it. See Section 11.2. (ii) The lines are circles in a clockwise direction if you are looking in the direction in which the current is travelling. (iii) If a.c. were used then the pattern would be changing continuously so fast that no field pattern would be observed. The compasses could not respond quickly enough.

   (b) (i) N-pole. (ii) Draw a diagram with lines parallel but curved near the edges of the magnet; direction C to B.

   (c) (i) When current is excessive the electromagnet attracts B and the circuit is broken. (ii) (1) A *much* larger current would be needed before the circuit breaker operated. (2) It would still work.

2. (a) (i) There is no change in flux, so no induced e.m.f. (ii) and (iii) There is a change of flux and, hence, an induced e.m.f. (see Sections 11.3 and 11.4).

   (b) See Example 11.3.

   (c) See Section 11.5, and Examples 11.4, 11.5 and 11.6. (i) 3 V. (ii) 160 A. (iii) See Example 11.4. (iv) Such high currents are used in welding and induction furnaces.

3. (a) (i) See Section 10.2. Current = 4 A. (ii) See Section 11.5. Ratio is 40 : 1. (iii) See Section 11.5 and Example 11.6. Current = 0.1 A. (iv) See Example 11.4.

   (b) (i) See Section 10.2. Resistance = 1.5 Ω. (ii) $(240 - 6)$ V = 234 V. (iii) See Section 10.1(a). Resistance = 58.5 Ω. (iv) See Section 10.2. Energy dissipated = 936 W.

   (c) Less energy lost as heat.

4. (a) and (b) See Section 11.5 and Example 11.4.

   (c) (i) 0.8 V. (ii) 2 V. (iii) 4 turns. (iv) 10 turns.

5. (a) (i) A greater deflection to the right. (ii) It will deflect to the right. (iii) No deflection. [There is no changing flux and no lines of flux are cutting the wire, so there is no induced e.m.f.]

   (b) (i) It will decrease. (ii) There will be little change. Hardly any current will pass through the high resistance in parallel with a low-resistance galvanometer.

   (c) The sides of a coil rotating in a magnetic field are cutting lines of flux. When a conductor cuts lines of flux, there is an induced e.m.f. If the circuit is closed, this e.m.f. produces an electric current. It is an a.c. current, because every half-revolution the sides of the coil cut the flux in the opposite direction.

6. (a) Iron oxide is easily magnetised and keeps its magnetism. The coating must be on the side closest to the recording head.

(b) Iron. It is easily magnetised, but it loses its magnetism quickly when the current is removed. The strength of the magnetism changes quickly with a changing current.

(c) The changing current in the wire results in a changing magnetic field in the iron, and this magnetic field magnetises the tape.

(d) Put the tape closer to the electromagnet, put the poles of the electromagnet closer together, put more turns of wire round the electromagnet.

(e) As the varying magnetic field of the tape passes the iron, an e.m.f. is induced in the wire. The resulting current passing through the wire reproduces the original alternating current which magnetised the tape.

(f) In (45×60) s, a length of tape $(4.7 \times 45 \times 60)$ cm $= 127$ m must pass through the tape recorder.

(g) Make the tape thinner, but it will not be so strong and will break more easily.

12 Electrostatics, the Cathode Ray Oscilloscope and Radioactivity

12.1 Electrostatics

There are two sorts of charge, positive and negative. A polythene rod rubbed with wool becomes negatively charged, and a cellulose acetate rod rubbed with wool becomes positively charged. Like charges repel and unlike charges attract (see Example 12.1).

A negatively charged polythene rod attracts an uncharged piece of paper, because the near side of the paper becomes positively charged, owing to the presence of the rod. This positive charge is attracted by the negative charge on the rod.

The magnitude of the force between charges decreases rapidly as the distance between them is increased.

One way of removing pollution in the form of smoke (dust particles) from the atmosphere is by electrostatic precipitation. The smoke is passed through a strong electric field, which produces ions (an atom that has gained or lost one or more electrons, and is therefore a charged atom) which adhere to the dust particles, giving them a charge. The charged particles are attracted towards an earthed plate and collect on it. They are periodically removed by striking the plate with a mechanical hammer. The dust particles fall into collectors for disposal.

12.2 Beams of Electrons

A heated filament emits electrons from the surface of the metal (thermionic emission). A beam of electrons may be produced by positioning a positively charged anode with a small hole in it, close to the filament. In a *cathode ray oscilloscope (CRO)*, the beam is passed through X-plates which produce an electric field that deflects the beam horizontally and Y-plates which produce an electric field that deflects it vertically. The input terminals connect the applied potential difference across the Y-plates. When the beam strikes the fluorescent screen at the end of the tube a spot of light is produced. When an alternating current is connected across the Y-plates, a vertical line appears on the screen. The waveform is displayed if the timebase (which moves the spot across the screen horizontally) is then switched on (see Examples 12.2 and 12.3).

In a TV tube an electron beam 'scans' the screen — that is, moves across the screen in a series of lines from left to right. The greater the number of electrons that strike any one spot on the screen, the brighter the picture at that point.

12.3 Radioactivity

(a) Origin and Detection

Radioactivity originates in the nucleus of the atom and is a random process. Radioactive substances have the ability to ionise the air surrounding them and it is this property which is used to detect them. In a cloud chamber alcohol vapour condenses along the path of the ionising radiation. In a Geiger–Müller tube, a pulse of current, produced by the ionising particle entering the tube, flows between the central electrode and the metal tube surrounding it. These pulses may be counted by a ratemeter or a scaler.

The correct count rate is obtained by subtracting the background count rate from the recorded count rate.

(b) α-, β- and γ-radiation

(i) α-radiation

This is the emission of positively charged helium nuclei from the nucleus of an atom. It is stopped by a sheet of paper. It can be deflected by electric and magnetic fields, but the experiment cannot be conducted in air because the path of an α-particle in air is only about 5 cm. α-radiation is the most highly ionising of the three radiations.

(ii) β-radiation

This is the emission of electrons from the nucleus of the atom. It is absorbed by about 2 mm of aluminium foil. It is easily deflected by electric and magnetic fields and is less ionising than α-particles.

(iii) γ-radiation

This is the emission of electromagnetic waves from the nucleus of an atom. It is unaffected by electric and magnetic fields. It is very penetrating but about 4 cm of lead will absorb most of it. It is only weakly ionising.

(c) Law of Decay

The law of radioactive decay states that the rate of radioactive decay at any instant is proportional to the quantity of radioactive material present at that instant.

(d) Half-life

The half-life of a radioactive source is the time it takes for the activity of the source to fall to half its original value (irrespective of what this value may be).

It may be measured by having a G–M tube connected to a ratemeter positioned at a suitable distance from the source. The count rate is plotted against time (to get the correct count rate, the background count rate must be subtracted from the reading on the ratemeter). The half-life is calculated from the graph (see Example 12.12).

(e) The Structure of the Atom, Proton Number, Nucleon Number

The positive charge of the atom and most of its mass is concentrated into a very small volume within the atom, called the nucleus. Electrons, spread over a large volume, surround the nucleus. The nucleus consists of positively charged protons, and uncharged particles of almost the same mass as a proton, called neutrons. In a neutral atom the number of electrons is equal to the number of protons. An electron has a mass approximately $\frac{1}{1800}$ of the mass of a proton or a neutron.

Proton number (atomic number) = number of protons in nucleus
Nucleon number (mass number) = number of protons in nucleus + number of neutrons in nucleus

All atoms of the same element have the same number of protons in the nucleus and hence contain the same number of orbiting electrons.

(f) Isotopes

Isotopes are atoms of a given element which differ only in the number of neutrons in the nucleus. Isotopes have the same proton number but a different nucleon number. They have the same number of electrons surrounding the nucleus and, hence, the same chemical properties.

(g) Equations for Radioactive Decay

$^{238}_{92}$U decays by α($^{4}_{2}$He) emission into thorium. The equation representing the decay is

$$^{238}_{92}\text{U} \xrightarrow[\text{emission}]{\alpha} {}^{234}_{90}\text{Th} + {}^{4}_{2}\text{He}$$

(The sum of the nucleon numbers on the right-hand side of the equation is equal to 238, and the sum of the proton numbers is 92.)

$^{239}_{92}$U decays by β($^{0}_{-1}$e) emission into neptunium. The equation representing the decay is

$$^{239}_{92}\text{U} \xrightarrow[\text{emission}]{\beta} {}^{239}_{93}\text{Np} + {}^{0}_{-1}\text{e}$$

(h) Some Uses of Radioactivity

(i) *Tracers*

A small quantity of a radioactive isotope is mixed with a non-radioactive isotope, enabling the path of the element to be followed in a plant or animal or human.

(ii) *Sterilisation*

γ-rays can be used to kill bacteria on such things as hospital blankets and certain foods.

(iii) *Thickness Control*

In paper mills the thickness of paper can be controlled by measuring how much β-radiation is absorbed by the paper. The source is placed on one side of the paper and the detector on the other side (see Examples 12.4 and 12.7).

(iv) *Radiotherapy*

The controlled use of γ-rays may be used to kill malignant cancer cells.

(i) Safety

Radiation is dangerous because it can harm living cells. The greatest risk comes from swallowing minute traces and therefore sources should never be held in the hand and no eating or drinking should take place near a radioactive source. Other precautions are (i) cover any cuts or sores, (ii) keep sources at a safe distance by holding them with long tweezers, (iii) never point the sources towards the human body, (iv) when not in use, keep the sources in their lead-lined boxes and (v) ensure that the exposure time to any radiation is kept to a minimum.

(j) Nuclear Power

In a nuclear power station nuclei of ^{235}U (or sometimes ^{239}Pu) each absorb a neutron and split into several parts, with the production of energy and two or three more neutrons. These neutrons may penetrate other ^{235}U nuclei, thus releasing more neutrons and starting a chain reaction.

Slow-moving neutrons are more easily absorbed than fast-moving ones, and moderators (usually water or graphite) are used to slow the neutrons down. The heat is extracted by means of a 'heat extractor fluid' which flows through the reactor and absorbs heat. A heat exchanger converts the heat into steam and the steam is used to drive the turbine generators.

The speed of the chain reaction is controlled by the use of control rods which absorb neutrons. When these rods are lowered into the reactor, the speed of the reaction is slowed down.

The energy produced comes from the conversion of a small amount of matter into energy.

An atomic bomb is an uncontrolled chain reaction.

Safety precautions in British nuclear power stations include:
(i) Regular monitoring of all personnel to check their exposure to radioactivity.

(ii) Careful disposal of all radioactive waste products (especially those with a long half-life).

(iii) An automatic shutdown of the reactor should the temperature of the reactor go above a certain safe limit.

(iv) Monitors throughout the whole plant to ensure that it is functioning properly and that there is no release of radioactive materials. This includes the monitoring of any water or gases which are discharged into the outside world (they are filtered before discharge).

(v) Barriers throughout the plant to prevent the release of radioactive material.

(vi) The wearing of protective clothing.

(vii) A complete shutdown every two years for a complete overhaul.

The Sun's power comes from *fusion*. The fusion of hydrogen-2 (deuterium) with hydrogen-3 (tritium) results in the release of energy. The hydrogen bomb is an uncontrolled fusion reaction. The technical problems of producing energy by a controlled fusion reaction have not yet been solved. One problem is that all ordinary containers melt at the temperature of the reaction.

12.4 Worked Examples

Example 12.1

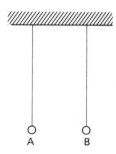

(a) Two small balls coated with metallic paint are suspended by insulating threads as shown in the diagram. Describe what you would observe if both balls were given a positive charge. How does the effect depend on the distance AB? **(3 marks)**

(b) Draw a diagram showing all the forces acting on one of the balls. **(3 marks)**

(c) In order to reduce the noise level of his machinery a manufacturer of nylon thread put rubber matting on the floor underneath his machinery. Small bits of fluff were found to stick to the nylon thread. Explain the reason for this. Someone suggested that the problem could be overcome by installing a humidifier to keep the air moist. Explain whether you think this was a reasonable suggestion. **(5 marks)**

Solution 12.1

(a) The balls would repel each other and both threads would be at an angle to the vertical. The smaller the distance between the threads the greater is the angle the threads make with the vertical.

(b)

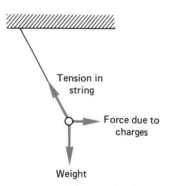

(c) Friction during the manufacture of the thread will cause the thread to become charged. If the charge on the thread is negative bits of fluff close to it will have a positive charge induced on the side nearest to the thread. The unlike charges attract and the fluff sticks to the thread.

Dry air is a good insulator but damp air is quite a good conductor. Moist air in contact with the thread would enable the charge to leak away, so the suggestion is a reasonable one.

Example 12.2

The diagram shows a waveform displayed on the screen of a cathode ray oscilloscope when an a.c. voltage is connected across the Y-plates. The Y gain is set at 2 V/cm and the time-base at 5 ms/cm. The graticule has a 1 cm grid.

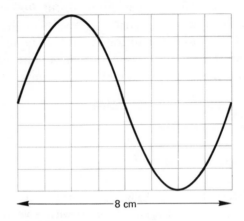

←——————— 8 cm ———————→

(a) What is the peak voltage of the waveform?
(b) What is the frequency of the a.c. voltage?
(c) Sketch on the graticule below the trace that would be seen on the screen if
    (i) the timebase setting were changed to 10 ms/cm. Label your trace A.
    (ii) the timebase setting were left at 5 ms/cm but the frequency of the applied voltage were halved. Label your trace B.

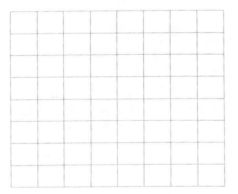

Solution 12.2

(a) $2 \dfrac{V}{cm} \times 4 \text{ cm} = 8 \text{ V}$.

(b) One cycle traces out 8 cm on the graticule and, hence, it takes

$$5 \frac{ms}{cm} \times 8 \text{ cm} = 40 \text{ ms} = \frac{40}{1000} \text{ s} = \frac{1}{25} \text{ s}$$

If each cycle takes $\dfrac{1}{25}$ s, then it happens 25 times every second.

Frequency = 25 Hz

180

(c) [If the timebase takes 10 ms instead of 5 ms to go 1 cm, it is travelling at half the speed. Twice as many cycles will appear on the screen. If the timebase is left at 5 ms/cm and the frequency is halved, then in the time it takes the spot to cross the screen there will be half a cycle on the screen.]

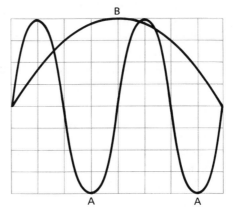

Example 12.3

Sketch the pictures seen on the screen of a cathode ray oscilloscope when the oscilloscope is adjusted so that the spot is in the middle of the screen and
(a) a battery is connected across the Y-plates. **(2 marks)**
(b) the output terminals from a transformer connected to the mains is connected across the Y-plates. **(2 marks)**
(c) the transformer is left connected to the Y-plates and the timebase is switched on. **(2 marks)**
(d) As in (c) but the speed of the timebase is doubled. **(2 marks)**

Solution 12.3

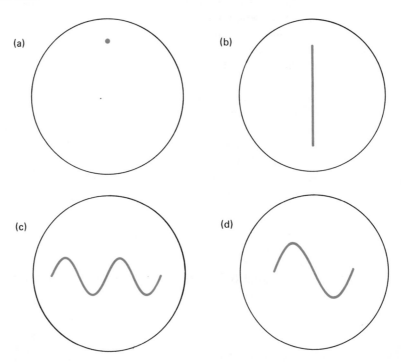

[In (b) the spot will be moving up and down many times a second, so the trace will appear as a straight line. In (d) the timebase is moving twice as fast, so that the oscillation will only have time to go through half the number of oscillations.]

Example 12.4

(a) What is meant by radioactivity?   (3 marks)
(b) Which type of radiation is
     (i) the most penetrating?
     (ii) the most ionising?
     (iii) used to sterilise medical instruments?
     (iv) absorbed by a piece of paper?   (4 marks)
(c) Give one example of the use of β radiation.   (4 marks)

Solution 12.4

(a) Radioactivity is the emission of α, β or γ radiation from the nucleus of unstable nuclei.
(b) (i) γ radiation; (ii) α radiation; (iii) γ radiation; (iv) α radiation.
(c) β radiation is used to monitor the thickness of paper being produced in a paper mill. The source is placed on one side of the paper and a detector (for example, a G–M tube connected to a ratemeter) on the other side of the paper. An increase in the count rate would mean that the thickness of the paper had decreased. The reading from the G–M tube is used to adjust the rollers which determine the thickness of the paper.

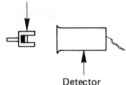

Radioactive source

Detector

Examples 12.5 and 12.6

A detector of radioactivity is placed opposite a radioactive source as shown in the diagram. It is found that radiation from the source is detected only if the distance between the source and the detector is less than 5 cm.
12.5   The source is probably emitting
A alpha radiation only
B beta radiation only
C gamma radiation only
D alpha and beta radiation only
E alpha, beta and gamma radiation
12.6   The source is changed. A sheet of paper placed between the new source and the detector is found to reduce the count rate. A sheet of aluminium is found to stop it completely. The source is emitting
A gamma radiation only
B beta and gamma radiation only
C alpha and gamma radiation only
D alpha and beta radiation only
E alpha, beta and gamma radiation   (L)

Solution 12.5

[β and γ radiation travel much more than 5 cm in air.]

<u>Answer A</u>

Solution 12.6

[αs would be stopped by a sheet of paper and βs by the sheet of aluminium. γ rays would penetrate the aluminium.]

<u>Answer D</u>

Example 12.7

The thickness of sheet metal produced at a steel mill is often monitored by means of *radio-activity*. A source is placed on one side of the sheet and a detector connected to a counter on the other side of the sheet. As the sheet passes along the production line, the readings on the counter over a period of 14 s are shown below.

Time/s	0	2	4	6	8	10	12	14
Total count	0	60	120	180	250	320	380	440
Count in 2 s	−	60	60	60				

(a) Would an α, β or γ source be the best type of source? Explain why the other two types of source would not be so suitable. **(3 marks)**

(b) Explain the meaning of the word 'radioactivity'. **(3 marks)**

(c) Complete the table shown above and explain what is happening to the thickness of the foil during the 14 s period. **(6 marks)**

(d) If the sheet metal got *very* thick, might the count rate fall to zero? Explain your answer. **(2 marks)**

Solution 12.7

(a) A β source would be best. α radiation would not penetrate the steel and the variation in γ radiation with small changes in thickness would be very small and not as easy to detect as the variation when a β source is used.

(b) The word 'radioactive' is used to describe the emission from the nucleus of an unstable atom of α, β or γ radiation.

(c)

Time/s	0	2	4	6	8	10	12	14
Total count	0	60	120	180	250	320	380	440
Count in 2 s	−	60	60	60	70	70	60	60

After 6 s and before 8 s the thickness has started to decrease. By 12 s the decrease has been corrected, and the thickness from 12 s to 14 s is the same as in the first 6 s.

(d) The count rate would never fall to zero, because of the background count that would always be present.

Example 12.8

The element X has an atomic mass number of 238 and an atomic number of 92. It emits an alpha particle forming an element Y. Y can be represented by

A $^{234}_{90}Y$ **B** $^{236}_{90}Y$ **C** $^{235}_{91}Y$ **D** $^{238}_{92}Y$ **E** $^{238}_{93}Y$ (AEB)

Solution 12.8

[See Section 12.3(g).]

Answer **A**

Example 12.9

An element P has an atomic mass of 239 and an atomic number of 92. It emits a β-particle forming an element Q. Q can be represented by

A $^{239}_{91}Q$ B $^{239}_{92}Q$ C $^{239}_{93}Q$ D $^{238}_{92}Q$ E $^{235}_{90}Q$

Solution 12.9

[See Section 12.3(g).]

Answer **C**

Example 12.10

The half-life of a radioactive substance is 10 hours. The original activity of a sample is measured and found to be 1200 counts per minute. Which of the following statements is correct?
A After 10 hours the count rate will be 120 counts per minute
B After 20 hours the count rate will be 300 counts per minute
C After 40 hours the count rate will be 150 counts per minute
D After 50 hours the count rate will be 240 counts per minute

Solution 12.10

[The activity halves every 10 hours. So
after 10 hours the activity will be 600 counts per minute
after 20 hours the activity will be 300 counts per minute
after 30 hours the activity will be 150 counts per minute
after 40 hours the activity will be 75 counts per minute.]

Answer **B**

Example 12.11

PART I
Start a new page

(a) Name *three* types of radiation emitted by radioactive sources. **(1 mark)**
(b) State, justifying your choice in each case, which of the radiations named in (a)
   (i) carries a negative charge
   (ii) is similar to X-rays
   (iii) is most easily absorbed
   (iv) travels with the greatest speed
   (v) is not deflected by a magnetic field
   (vi) is emitted when $^{238}_{92}U$ decays to $^{234}_{90}Th$
   (vii) is similar in nature to cathode rays **(7 marks)**

(c) Carbon-14 ($^{14}_{6}$C) is an *isotope* of carbon. It is radioactive, decaying to nitrogen-14 ($^{14}_{7}$N).
  (i) What is the meaning of the term *isotope*?
  (ii) Write an equation for the decay of carbon-14. **(3 marks)**

(d) Carbon-14 has a half-life of 5600 years.
  (i) What is the meaning of the term *half-life*?
  (ii) Draw a graph to show the decay of carbon-14 from an initial activity of 64 counts per minute. **(7 marks)**

(e) While trees and plants live they absorb and emit carbon-14 (in the form of carbon dioxide) so that the amount of the isotope remains constant.
  (i) What happens to the amount of carbon-14 after a tree dies?
  (ii) A sample of wood from an ancient dwelling gives 36 counts per minute. A similar sample of living wood gives 64 counts per minute. From your graph deduce the approximate age of the dwelling. **(3 marks)**

(AEB)

Solution 12.11

(a) Alpha-, beta- and gamma-radiation.

(b) (i) β-radiation is normally negatively charged (although positive β-particles are sometimes emitted).
  (ii) γ-radiation and X-rays are both electromagnetic radiation.
  (iii) α-radiation will not pass through a sheet of paper. The others easily penetrate a piece of paper.
  (iv) γ-radiation like all electromagnetic radiation travels at the speed of light.
  (v) γ-radiation. Both α- and β-radiation are charged and deflected by a magnetic field.
  (vi) α-radiation. It is a helium nucleus $^{4}_{2}$He.
  (vii) β-radiation. Cathode rays and β-radiation are both fast moving electrons.

(c) (i) Isotopes are atoms of a given element which differ only in the number of neutrons in the nucleus. Isotopes therefore have the same proton number but a different nucleon number.
  (ii) $^{14}_{6}$C $\rightarrow$ $^{14}_{7}$N + $^{0}_{-1}$e

(d) (i) The half-life is the time for the activity to halve. Thus in 5600 years the activity will halve. In 11 200 years it will have become a quarter of its original value.

(ii) Counts/min^{-1}

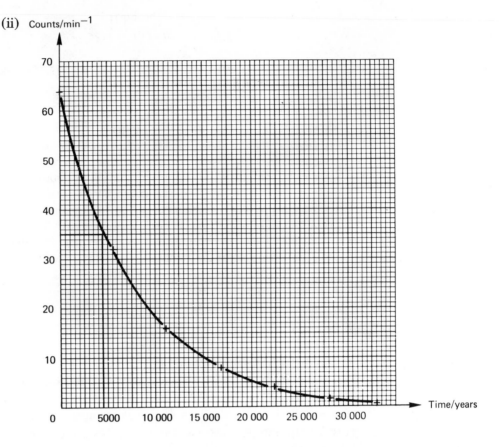

Time/years

(e) (i) The carbon-14 decays and every 5600 years the activity will halve.
     (ii) 4500 years.

Example 12.12

In an experiment to determine the half-life of radon-220 ($^{220}_{86}$Rn) the following results were obtained, after allowing for the background count:

Time/s	0	10	20	30	40	50	60	70
Count rate/s^{-1}	30	26	23	21	18	16	14	12

(a) By plotting the count rate (vertically) against the time (horizontally), determine the half-life of $^{220}_{86}$Rn. Show clearly on your graph how you obtain your answer. **(6 marks)**

(b) (i) What is the origin of the background count? **(2 marks)**
     (ii) How is the background count determined? **(3 marks)**

(c) $^{220}_{86}$Rn emits α-particles.
     (i) What is an α-particle? **(2 marks)**
     (ii) When $^{220}_{86}$Rn emits an α-particle it becomes an isotope of the element polonium (Po). Write an equation to represent this change. **(2 marks)**

(d) When carrying out experiments with radioactive sources, students are instructed that
     (i) the source should never be held close to the human body,
     (ii) no eating or drinking is allowed in the laboratory.
     Why is it important to follow these instructions? **(5 marks)**
     (L)

Solution 12.12

(a) [The half-life is the time for the count rate to halve no matter where it starts. You should always take a number of different starting points and take the average value for the half-life. In this case only two were taken because the

186

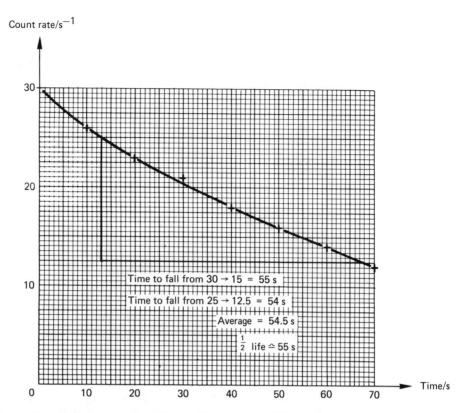

Count rate/s^{-1}

Time to fall from 30 → 15 = 55 s

Time to fall from 25 → 12.5 = 54 s

Average = 54.5 s

$\frac{1}{2}$ life ≏ 55 s

Time/s

readings stopped soon after the count rate halved. Normally at least three values would be calculated.]

(b) (i) The background count is always present in the atmosphere. It arises from cosmic rays entering the atmosphere, radioactive materials on the surface of the Earth, radon in the atmosphere and X-rays from television screens.

(ii) In the above experiment the detector is removed from the radon source and left in the atmosphere. The count is taken over a period of say a minute and the count rate per second determined by dividing by 60. The average of about 10 counts would be taken.

(c) (i) An α-particle is a helium nucleus, $_2^4$He. It has a charge of +2 and a nucleon number of 4.

(ii) $_{86}^{220}$Rn = $_{84}^{216}$Po + $_2^4$He

(d) Radiation is dangerous because it can harm living cells.

(i) The intensity of radiation falls off rapidly as the distance increases, so the further the source is from the human body the less is the risk.

(ii) α-radiation cannot penetrate the skin so the greatest risk comes from swallowing minute traces. Indeed any radiation is more dangerous if it is inside the body; hence, eating and drinking should never take place in the vicinity of radioactive materials.

Harm to the body could result if these instructions were not followed.

Example 12.13

A student suggests that a stable isotope of an element X should be represented by the form $_3^7$X. Another student suggests the form $_3^4$X for another stable isotope.

(a) State why these two forms represent isotopes of the element X.

(b) How many neutrons are there in an atom of
(i) $_3^4$X, (ii) $_3^7$X?

(c) Which is likely to be the form of the stable isotope? Give a reason for your answer.

(6 marks)

(L)

Solution 12.13

(a) They have the same proton number, 3, and hence the same number of protons, the same number of electrons and the same chemical properties. They have different numbers of neutrons in the nucleus.

(b) (i) 1, (ii) 4. [(4 − 3) and (7 − 3).]

(c) Light elements are stable if the numbers of protons and neutrons are approximately equal. ^7_3X is likely to be more stable because it has 3 protons and 4 neutrons.

Example 12.14

In a nuclear reactor the following reaction takes place

$$^{235}_{92}U + ^1_0n \rightarrow ^{236}_{92}U$$

The $^{236}_{92}U$ formed is unstable and disintegrates, with the release of two or three neutrons, together with the release of a considerable amount of energy.

(a) What do the numbers 236 and 92 represent? **(2 marks)**

(b) Explain the meaning of the term 'isotope' with reference to $^{235}_{92}U$ and $^{236}_{92}U$. **(3 marks)**

(c) Use the information given above to explain what is meant by a chain reaction. **(3 marks)**

(d) What are moderators? Why are they necessary in a nuclear reactor? **(3 marks)**

(e) What is the source of energy in the above reaction? **(2 marks)**

(f) Explain how the heat is removed from the core of a reactor and how this heat is used to generate electricity. **(3 marks)**

(g) Many nuclear waste products have long half-lives. Explain the term 'half-life' and discuss why the long half-life of waste products presents a health hazard. **(4 marks)**

Solution 12.14

(a) 236 is the nucleon number and represents the number of neutrons plus the number of protons in the nucleus. 92 is the proton number and is the number of protons in the nucleus.

(b) Isotopes are atoms of a given element which differ only in the number of neutrons in the nucleus. Isotopes have the same number of protons in the nucleus (92 in $^{235}_{92}U$ and $^{236}_{92}U$), but a different number of neutrons: (235 − 92) = 143 in $^{235}_{92}U$ and (236 − 92) = 144 in $^{236}_{92}U$.

(c) If the two or three neutrons released when $^{236}_{92}U$ disintegrates are captured by other $^{235}_{92}U$ atoms, then more neutrons are released. As the process continues, more and more neutrons are released. A chain reaction has started.

(d) Moderators are used to reduce the speed of released neutrons. Slow-moving neutrons are more easily captured by $^{235}_{92}U$ than fast-moving ones.

(e) The source of energy is the reduction of the mass of the atom. Some of the mass is turned into energy.

(f) A heat extractor fluid flows around the reactor and the fluid's temperature rises as it absorbs heat from the reactor. A heat extractor converts this heat into steam; the steam is used to drive a turbine generator.

(g) The half-life is the time for the activity of a radioactive substance to fall to half its value. If the half-life is 1 h, after 1 h the activity is halved, after 2 h it is a quarter and after 3 h it is an eighth, so after a very short time the activity is reduced to a very small amount. On the other hand, if the half-life is 1000 years, then the activity remains at a very high level for a very long time and this can result in a health hazard.

12.5 Have You Mastered the Basics?

1. Can you describe how to use a CRO to (a) measure potential difference in d.c. and a.c. circuits, (b) display waveforms and (c) measure frequency?
2. Can you state the nature, charge and properties of α-, β- and γ-radiation?
3. Can you state the law of radioactive decay and explain the meaning of half-life?
4. Can you explain the meaning of proton number, nucleon number and isotope?
5. Can you describe some uses of radioactivity and the safety precautions necessary when using them?
6. Can you write equations to illustrate α and β decay?
7. Do you understand the action of a nuclear reactor?
8. The half-life of a radioactive isotope which emits β particles is 24 days. It has a mass of 1 kg and its activity is found to be 2000 counts/min.
   (a) What is the activity after (i) 48 days and (ii) 96 days?
   (b) Has its mass changed significantly after 96 days?

12.6 Answers and Hints on Solutions to 'Have You Mastered the Basics?'

1. See Section 12.2 and Examples 12.2 and 12.3.
2. See Section 12.3(b).
3. See Section 12.3(c) and (d). Also Examples 12.10 and 12.12.
4. See Section 12.3(e).
5. See Section 12.3(h) and (i).
6. See Section 12.3(g) and Examples 12.8, 12.9 and 12.12.
7. See Section 12.3(j) and Example 12.14.
8. After 24 days the count rate is 1000 counts/min.
   After 48 days the count rate is 500 counts/min.
   After 72 days the count rate is 250 counts/min.
   After 96 days the count rate is 125 counts/min.
   (a) (i) 500 counts/min. (ii) 125 counts/min.
   (b) No. The mass of a β particle is *very* small, so the mass of the isotope has not changed appreciably. However, the mass of the undecayed isotope is 0.25 kg and 0.75 kg of the isotope has decayed.

12.7 Questions

(Answers and hints on solutions will be found in Section 12.8.)

Question 12.1

The spot on a CRO is stationary in the centre of the screen. What adjustments have been made to the CRO and what has been connected across the input terminals to obtain each of the traces shown in (i), (ii) and (iii) in the diagrams below?

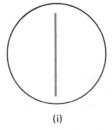

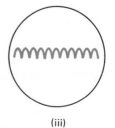

(i)   (ii)   (iii)

Question 12.2

(a) How may a cathode-ray oscilloscope be used to measure
  (i) the e.m.f. of a rechargeable cell
  (ii) the maximum or peak value of the e.m.f. of a dynamo which is rotating at constant speed. **(8 marks)**
(b) The timebase of an oscilloscope is switched on and the switch set to 5 milliseconds per centimetre. The distance across the screen is 12 cm.
  Sketch the appearance of the trace on the screen, if an oscillation of frequency 50 Hz is fed to the input. **(6 marks)**

Question 12.3

Which of the following statements is true? Isotopes of any one element
A have the same chemical properties
B contain the same number of neutrons and protons
C contain different numbers of protons in the nucleus
D contain the same number of electrons and neutrons

Question 12.4

The symbol $^{235}_{92}U$ denotes an *isotope* of the element uranium of *nucleon number* 235 and *proton number* 92. Explain the meaning of the terms in italics. **(6 marks)**
  State the changes which take place in the nucleus of a radioactive element when an alpha particle is emitted. **(4 marks)**
  Hence show, using the above symbol notation, the change when radium ($^{226}_{88}Ra$) emits an alpha particle and radon (Rn) is formed. **(2 marks)**
  State *two* uses of radioactive isotopes and *one* precaution necessary when using such substances. **(3 marks)**
  (SUJB)

Question 12.5

An isotope is said to have a half-life of 6 years. This means that
A After 1 year $\frac{1}{6}$ of the isotope has decayed
B After 1 year $\frac{1}{6}$ of the isotope remains undecayed
C After 12 years $\frac{1}{4}$ of the isotope has decayed
D After 12 years $\frac{1}{4}$ of the isotope remains undecayed
E After 12 years the isotope has completely decayed

Question 12.6

(a) Compile a table for α, β, and γ radiation which shows (i) their nature, (ii) their relative penetrating powers, (iii) their ionising powers. **(8 marks)**
(b) You are asked to determine the level of shaving cream in an aerosol. Explain how you would do this using a radioactive source and a detector. **(6 marks)**
(c) The half-life of an element is 8 days. How much of it will be left after 24 days? **(6 marks)**

Question 12.7

What is meant by the half-life of a radioactive element? Use the table of readings below to calculate the half-life of the element.

Time/s	0	110	163
Count rate/min^{-1}	2000	500	250

After what time would you expect the count rate to be 125?

Question 12.8

(a) A Geiger–Müller tube attached to a scaler is placed on a bench in the laboratory. Over three consecutive minutes the scaler reads 11, 9 and 16 counts per minute.

When a radioactive source is placed near to the Geiger–Müller tube the counts over three consecutive minutes are 1310, 1270, and 1296 per minute.

When a piece of thick paper is placed between the source and the tube the counts are 1250, 1242, and 1236 per minute.

When the paper is replaced by a sheet of aluminium 2 mm thick the counts are 13, 12 and 11 per minute.

  (i) Why is there a reading when no source is present? **(2 marks)**
 (ii) Why do the three readings in any one group differ? **(2 marks)**
(iii) What can be deduced about the nature of the emission? Give reasons for your answer. **(5 marks)**

(b) What do you understand by the *half-life* of a radioactive element? **(2 marks)**

The graph below is plotted from readings taken with a radioactive source at daily intervals.

Use the graph to deduce the half-life of the source. **(2 marks)**

Hence give the count rate after five days, and the time when the count is 160 per minute. **(4 marks)**

Would you expect the mass of the source to have changed significantly after 4 days? (Give a reason for your answer.) **(3 marks)**

(L)

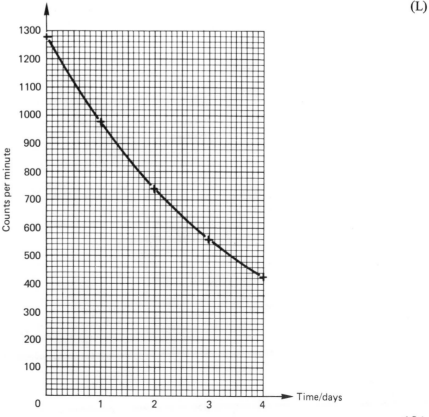

12.8 Answers and Hints on Solutions to Questions

1. See Sections 12.2 and 10.4(g) and Examples 12.2 and 12.3.

2. (a) (i) Set the spot in the centre of the screen and connect the cell across the Y-plates (marked 'input'). The knob marked 'volts/cm' should be set at about 1 volt/cm if two or three 1.5 V cells are used; for a greater voltage a different setting would be needed. (ii) A suitable setting is chosen for the volts/cm and the output terminals of the dynamo connected to the input terminals of the oscilloscope.

   (b) One cycle will occupy 4 cm of screen.

3. See Section 12.3(f).
   Answer **A**

4. See Section 12.3(e), (f) and (g). $^{222}_{86}$Rn is formed. See Section 12.3(h) and (i).

5. See Section 12.3(d), Example 12.10 and 'Have You Mastered the Basics?' No. 8.
   Answer **D**

6. (a) See Section 12.3(b).

   (b) Place a beta source on one side of the tin and a G–M tube on the other side directly opposite the source. Keep the source and detector in the same horizontal plane and move them up and down the outside of the tin.

   (c) See Example 12.10 and 'Have You Mastered the Basics?' No. 8.
   Answer $\frac{1}{8}$ of the element will be left

7. See Section 12.3(d) and Example 12.10. Two half-lives is 110 s; therefore one half-life is 55 s. One half-life is (163 − 110) s = 53 s. Average is 54 s. About 217 s.

8. (a) (i) See Example 12.12 and Section 12.3(d). (ii) Radioactivity is a random process. (iii) No significant reduction with paper; therefore not alpha. Radiation reduced to background count by 2 mm of aluminium and therefore not gamma. Must be beta.

   (b) See Section 12.3(d) and Example 12.10. About 2.5 days. After 5 days count rate is 320 counts per minute. The count rate will be 160 counts per minute after 7.5 days. The mass does not appreciably change when β-particles are emitted, because the mass of the emitted particles is small.

Some Basic Units

Unit and symbol	Quantity measured and usual symbol	Comments
second (s)	time (t)	The unit of time. 60 s in 1 minute.
metre (m)	length, distance (l, s)	Approximately the length of a good-sized stride.
kilometre (km)	length, distance (l, s)	1 km = 1000 m (a bit more than half a mile)
kilogram (kg)	mass (m)	The mass of the average bag of sugar is about 1 kg
newton (N)	force (F)	The pull of the Earth (weight) on an apple of average size is about 1 N.
pascal (Pa)	pressure (P)	1 Pa = 1 N/m^2. The pressure exerted when you push hard on a table with your thumb is about 1 million pascals.
joule (J)	energy (E)	1 J = 1 Nm. About the energy needed to place an apple of average size on a shelf 1 metre high. 4200 J (specific heat capacity) is needed to raise 1 kg of water through 1 K. About 2 million joules (specific latent heat) is needed to boil away 1 kg of water.
watt (W)	power (p)	1 W is a rate of working of 1 J/s. It is also the energy produced every second when 1 V causes 1 A to flow. Household mains lamps are usually between 40 W and 100 W. Watts = volts × amperes.
degree Celsius (°C)	temperature (t, θ)	The temperature of water changes by 100°C when going from the melting point to the boiling point.
kelvin (K)	temperature (T)	A temperature change of 1 kelvin is the same as a temperature change of 1 degree C.
hertz (Hz)	frequency (f)	1 Hz is one cycle per second. BBC radio broadcasts are about 1 MHz; VHF about 90 MHz.
ampere (A)	current (I)	The current in most torch and household bulbs is between 0.1 A and 0.4 A.
volt (V)	potential difference (V)	Many cassette players and torches use batteries which are 1.5 V. The mains voltage is 240 V.
ohm (Ω)	resistance (R)	A p.d. of 1 V across 1 Ω produces a current of 1 A. A torch bulb has a resistance of about 10 Ω. $V = IR$.
m/s	speed (v)	45 miles per hour is about 20 m/s. The speed of light is 3×10^8 m/s.
m/s^2	acceleration (a)	Objects falling on Earth accelerate at about 10 m/s^2. Family cars accelerate at about 2 m/s^2.

Index